REASONING WITH DATA

Reasoning with Data

An Introduction to Traditional
and Bayesian Statistics Using R

Jeffrey M. Stanton

THE GUILFORD PRESS
New York   London

Copyright © 2017 The Guilford Press
A Division of Guilford Publications, Inc.
370 Seventh Avenue, Suite 1200, New York, NY 10001
www.guilford.com

Printed in the United States of America

This book is printed on acid-free paper.

Last digit is print number: 9   8   7   6   5   4   3   2   1

Library of Congress Cataloging-in-Publication

Names: Stanton, Jeffrey M., 1961–
Title: Reasoning with data : an introduction to traditional and Bayesian
  statistics using R / Jeffrey M. Stanton.
Description: New York : The Guilford Press, [2017] | Includes bibliographical
  references and index.
Identifiers: LCCN 2017004984| ISBN 9781462530267 (pbk. : alk. paper) |
  ISBN 9781462530274 (hardcover : alk. paper)
Subjects: LCSH: Bayesian statistical decision theory—Problems, exercises,
  etc. | Bayesian statistical decision theory—Data processing. |
  Mathematical statistics—Problems, exercises, etc. | Mathematical
  statistics—Data processing. | R (Computer program language)
Classification: LCC QA279.5 .S745 2017 | DDC 519.50285/53—dc23
LC record available at *https://lccn.loc.gov/2017004984*

Preface

Back in my youth, when mammoths roamed the earth and I was learning statistics, I memorized statistical formulas and learned when to apply them. I studied the statistical rulebook and applied it to my own data, but I didn't fully grasp the underlying logic until I had to start teaching statistics myself. After the first couple semesters of teaching, and the first few dozen really confused students (sorry, folks!), I began to get the hang of it and I was able to explain ideas in class in a way that made the light bulbs shine. I used examples with data and used scenarios and graphs that I built from spreadsheets to illustrate how hypothesis testing really worked from the inside. I deemphasized the statistical formulas or broke them open so that students could access the important concepts hiding inside the symbols. Yet I could never find a textbook that complemented this teaching style—it was almost as if every textbook author wanted students to follow the same path they themselves had taken when first learning statistics.

So this book tries a new approach that puts simulations, hands-on examples, and conceptual reasoning first. That approach is made possible in part thanks to the widespread availability of the free and open-source R platform for data analysis and graphics (R Core Team, 2016). R is often cited as the language of the emerging area known as "data science" and is immensely popular with academic researchers, professional analysts, and learners. In this book I use R to generate graphs, data, simulations, and scenarios, and I provide all of the commands that teachers and students need to do the same themselves.

One definitely does not have to already be an R user or a programmer to use this book effectively. My examples start slowly, I introduce R commands and data structures gradually, and I keep the complexity of commands and code sequences to the minimum needed to explain and explore the statistical concepts. Those who go through the whole book will feel competent in using R and will have a lot of new problem-solving capabilities in their tool belts. I know this to be the case because I have taught semester-long classes using earlier drafts of this textbook, and my students have arrived at their final projects with substantial mastery of both statistical inference techniques and the use of R for data analysis.

Above all, in writing this book I've tried to make the process of learning data analysis and statistical concepts as engaging as possible, and possibly even fun. I wanted to do that because I believe that quantitative thinking and statistical reasoning are incredibly important skills and I want to make those skills accessible to a much wider range of people, not just those who must take a required statistics course. To minimize the "busy work" you need to do in order to teach or learn from this book, I've also set up a companion website with a copy of all the code as well as some data sets and other materials that can be used in- or outside of the classroom (*www.guilford.com/stanton2-materials*). So, off you go, have fun, and keep me posted on how you do.

In closing, I acknowledge with gratitude Leonard Katz, my graduate statistics instructor, who got me started on this journey. I would also like to thank the initially anonymous reviewers of the first draft, who provided extraordinarily helpful suggestions for improving the final version: Richard P. Deshon, Department of Psychology, Michigan State University; Diana Mindrila, Department of Educational Research, University of West Georgia; Russell G. Almond, Department of Educational Psychology, Florida State University; and Emily A. Butler, Department of Family Studies and Human Development, University of Arizona. Emily, in particular, astutely pointed out dozens of different spots where my prose was not as clear and complete as it needed to be. Note that I take full credit for any remaining errors in the book! I also want to give a shout out to the amazing team at Guilford Publications: Martin Coleman, Paul Gordon, CDeborah Laughton, Oliver Sharpe, Katherine Sommer, and Jeannie Tang. Finally, a note of thanks to my family for giving me the time to lurk in my basement office for the months it took to write this thing. Much obliged!

Contents

Contents

Purchasers of this book can find annotated R code
from the book's examples, in-class exercises,
links to online videos, and other resources
at *www.guilford.com/stanton2-materials.*

Introduction

William Sealy Gosset (1876–1937) was a 19th-century uber-geek in both math and chemistry. The latter expertise led the Guinness Brewery in Dublin, Ireland, to hire Gosset after college, but the former made Gosset a household name in the world of statistics. As a forward-looking business, the Guinness brewery was alert for ways of making batches of beer more consistent in quality. Gosset stepped in and developed what we now refer to as "small-sample statistical techniques"—ways of generalizing from the results of a relatively few observations (Lehmann, 2012).

Brewing a batch of beer takes time, and high-quality ingredients are not cheap, so in order to draw conclusions from experimental methods applied to just a few batches, Gosset had to figure out the role of chance in determining how a batch of beer had turned out. The brewery frowned upon academic publications, so Gosset had to publish his results under the modest pseudonym "Student." If you ever hear someone discussing the "Student's t-test," that is where the name came from.

The Student's t-test allows us to compare two groups on some measurement. This process sounds simple, but has some complications hidden in it. Let's consider an example as a way of illustrating both the simplicity and the complexity. Perhaps we want to ask the question of whether "ale yeast" or "lager yeast" produces the higher alcohol content in a batch of beer. Obviously, we need to brew at least one batch of each type, but every brewer knows that many factors influence the composition of a batch, so we should really brew several batches with lager yeast and several batches with ale yeast. Let's brew five batches of each type. When we are done we can *average* the measurements of alcohol content across the five batches made with lager yeast and do the same across the five batches made with ale yeast.

What if the results showed that the average alcohol content among ale yeast batches was slightly higher than among lager yeast batches? End of story, right? Unfortunately, not quite. Using the tools of mathematical probability available in the late 1800s, Gosset showed that the average only painted part of the big picture. What also mattered was how **variable** the batches were—in other words, Was there a large *spread* of results among the observations in either or both groups? If so, then one could not necessarily rely upon one observed difference between two averages to generalize to other batches. Repeating the experiment might easily lead to a different conclusion. Gosset invented the *t*-test to quantify this problem and provide researchers with the tools to decide whether any observed difference in two averages was sufficient to overcome the natural and expected effects of **sampling error.** Later in the book, I will discuss both sampling and sampling error so that you can make sense of these ideas.

Well, time went on, and the thinking that Gosset and other statisticians did about this kind of problem led to a widespread tradition in applied statistics known as **statistical significance testing.** "Statistical significance" is a technical term that statisticians use to quantify how likely a particular result might have been in light of a model that depicts a whole range of possible results. Together, we will unpack that very vague definition in detail throughout this book. During the 20th century, researchers in many different fields—from psychology to medicine to business—relied more and more on the idea of statistical significance as the most essential guide for judging the worth of their results. In fact, as applied statistics training became more and more common in universities across the world, lots of people forgot the details of exactly why the concept was developed in the first place, and they began to put a lot of faith in scientific results that did not always have a solid basis in sensible quantitative reasoning. Of additional concern, as matters have progressed we often find ourselves with so much data that the small-sample techniques developed in the 19th century sometimes do not seem relevant anymore. When you have hundreds of thousands, millions, or even billions of records, conventional tests of statistical significance can show many negligible results as being statistically significant, making these tests much less useful for decision making.

In this book, I explain the concept of statistical significance so that you can put it in perspective. Statistical significance still has a meaningful role to play in quantitative thinking, but it represents one tool among many in the quantitative reasoning toolbox. Understanding significance and its limitations will help you to make sense of reports and publications that you read, but will also help you grasp some of the more sophisticated techniques that we can now use to sharpen our reasoning about data. For example, many statisticians and researchers now advocate for so-called **Bayesian** inference, an approach to statistical reasoning that differs from the **frequentist** methods (e.g., statistical significance) described above. The term "Bayesian" comes from the 18th-century thinker Thomas Bayes, who figured out a fresh strategy for reasoning based on prior evidence. Once you have had a chance to digest all of these concepts and put them to work in your own examples, you will be in a position to critically examine other people's claims about data and to make your own arguments stronger.

GETTING STARTED

Contrary to popular belief, much of the essential material in applied statistics can be grasped without having a deep background in mathematics. In order to make sense of the ideas in this book, you will need to be able to:

- Add, subtract, multiply, and divide, preferably both on paper and with a calculator.
- Work with columns and rows of data, as one would typically find in a spreadsheet.
- Understand several types of basic graphs, such as bar charts and scatterplots.
- Follow the meaning and usage of algebraic equations such as $y = 2x-10$.
- Install and use new programs on a laptop or other personal computer.
- Write interpretations of what you find in data in your own words.

To illustrate concepts presented in this book as well as to practice essential data analysis skills, we will use the open-source R platform for statistical computing and graphics. R is powerful, flexible, and especially "extensible" (meaning that people can create new capabilities for it quite easily). R is free for everyone to use and it runs on just about every kind of computer. It is "command line" oriented, meaning that most of the work that one needs to perform is done through carefully crafted text instructions, many of which have "fiddly" syntax (the punctuation and related rules for writing a command that works). Sometimes it can be frustrating to work with R commands and to get the syntax just right. You will get past that frustration because you will have many working examples to follow and lots of practice at learning and using commands. There are also two appendices at the end of the book to help you with R. Appendix A gets you started with installing and using R. Appendix B shows how to manage the data sets you will be analyzing with R. Appendix C demonstrates a convenient method for sorting a data set, selecting variables, and calculating new variables. If you are using this book as part of a course, your instructor may assign these as readings for one of your initial classes or labs.

One of the virtues of R as a teaching tool is that it hides very little. The successful user must fully understand what is going on or else the R commands will not work. With a spreadsheet, it is easy to type in a lot of numbers and a formula like =FORECAST() and, bang, a number pops into a cell like magic, whether it makes any sense or not. With R you have to know your data, know what you can do with it, know how it has to be transformed, and know how to check for problems. The extensibility of R also means that volunteer programmers and statisticians are adding new capabilities all the time. Finally, the lessons one learns in working with R are almost universally applicable to other programs and environments. If one has mastered R, it is a relatively small step to get the hang of a commercial statistical system. Some of the concepts you learn in working with R will even be applicable to general-purpose programming languages like Python.

As an open-source program, R is created and maintained by a team of volunteers. The team stores the official versions of R at a website called CRAN—the Comprehensive R Archive Network (Hornik, 2012). If your computer has the Windows, Mac-OS-X, or Linux operating system, there is a version of R waiting for you at *http://cran.r-project.org*. If you have any difficulties installing or running the program, you will find dozens of great written and video tutorials on a variety of websites. See Appendix A if you need more help.

We will use many of the essential functions of R, such as adding, subtracting, multiplying, and dividing, right from the command line. Having some confidence in using R commands will help you later in the book when you have to solve problems on your own. More important, if you follow along with every code example in this book while you are reading, it will really help you understand the ideas in the text. This is a really important point that you should discuss with your instructor if you are using this book as part of a class: when you do your reading you should have a computer nearby so that you can run R commands whenever you see them in the text!

We will also use the aspect of R that makes it extensible, namely the "package" system. A *package* is a piece of software and/or data that downloads from the Internet and extends your basic installation of R. Each package provides new capabilities that you can use to understand your data. Just a short time ago, the package repository hit an incredible milestone—6,000 add-on packages—that illustrates the popularity and reach of this statistics platform. See if you can install a package yourself. First, install and run R as described just above or as detailed in Appendix A. Then type the following command at the command line:

```
install.packages("modeest")
```

This command fetches the "mode estimation" package from the Internet and stores it on your computer. Throughout the book, we will see R code and output represented as you see it in the line above. I rarely if ever show the command prompt that R puts at the beginning of each line, which is usually a " >" (greater than) character. Make sure to type the commands carefully, as a mistake may cause an unexpected result. Depending upon how you are viewing this book and your instructor's preferences, you may be able to cut and paste some commands into R. If you can cut and paste, and the command contains quote marks as in the example above, make sure they are "dumb" quotes and not "smart" quotes (dumb quotes go straight up and down and there is no difference between an open quote and a close quote). R chokes on smart quotes. R also chokes on some characters that are cut and pasted from PDF files.

When you install a new package, as you can do with the install.packages command above, you will see a set of messages on the R console screen showing the progress of the installation. Sometimes these screens will contain warnings. As long as there is no outright error shown in the output, most warnings can be safely ignored.

When the package is installed and you get a new command prompt, type:

```
library(modeest)
```

This command activates the package that was previously installed. The package becomes part of your active "library" of packages so that you can call on the functions that library contains. Throughout this book, we will depend heavily on your own sense of curiosity and your willingness to experiment. Fortunately, as an open-source software program, R is very friendly and hardy, so there is really no chance that you can break it. The more you play around with it and explore its capabilities, the more comfortable you will be when we hit the more complex stuff later in the book. So, take some time now, while we are in the easy phase, to get familiar with R. You can ask R to provide help by typing a question mark, followed by the name of a topic. For example, here's how to ask for help about the library() command:

```
?library
```

This command brings up a new window that contains the "official" information about R's library() function. For the moment, you may not find R's help very "helpful" because it is formatted in a way that is more useful for experts and less useful for beginners, but as you become more adept at using R, you will find more and more uses for it. Hang in there and keep experimenting!

## CHAPTER 1

Statistical Vocabulary

I'm the kind of person who thinks about the definitions of words and who tries to use the right word in the right situation. Knowing the correct terminology for important ideas makes it easier to think about and talk about things that really matter. As is the case with most fields, statistics has a specialized vocabulary that, when you learn it, helps you to think about and solve problems. In this chapter, I introduce some of the most essential statistical terminology and I explain how the concepts differ and are connected to one another. Once you have mastered this vocabulary, later parts of the book will be much easier to understand.

DESCRIPTIVE STATISTICS

Descriptive statistics accomplish just what you would think: they *describe* some **data.** (By the way, I typically use the word *data* as a plural noun because it usually refers to a collection of pieces of information.) Usually, descriptive statistics describe some data by summarizing them as a single number—what we call a **scalar** value. Most of the time, when we calculate a descriptive statistic we do so over a **vector** of numbers—that is, a list of scalar values that all quantify one attribute such as height, weight, or alcohol content. To put it all together, a **descriptive statistic** summarizes a vector of data by calculating a scalar value that summarizes those data. Let's look in detail at two kinds of descriptive statistics: measures of central tendency and measures of dispersion.

MEASURES OF CENTRAL TENDENCY

If you have ever talked about or calculated the **average** of a set of numbers, you have used a "measure of central tendency." Measures of central tendency summarize a vector of data by figuring out the location of the typical, the middle, or the most common point in the data. The three most common measures of central tendency are the mean, the median, and the mode.

The Mean

The most widely known statistic in the world is the **arithmetic mean,** or more commonly and simply, the **mean:** it is also the most commonly used measure of central tendency. When nonstatisticians use the term "average" they're almost always referring to the mean. As you probably know, the mean is computed by adding together all of the values in a vector—that is, by summing them up to compute a total, and then dividing the total by the number of observations in the vector. Let's say, for example, that we measured the number of centimeters of rainfall each day for a week. On Sunday, Monday, and Tuesday there was no rain. On Wednesday it rained 2 centimeters. On Thursday and Friday it rained 1 centimeter. Then Saturday was very drippy with 4 centimeters of rain. If we made a vector from these numbers, it would look like this:

$$0, 0, 0, 2, 1, 1, 4$$

You know what to do to get the mean: add these seven numbers together and then divide by 7. I get a sum of 8. Eight divided by 7 is about 1.14. That's the mean. Let's try this in R. At the command line, type the following commands. You do not have to type all of the stuff after the # character. Everything coming after the # character is a comment that I put in solely for the benefit of you, the human reader.

```
rainfall <- c(0,0,0,2,1,1,4)      # Amount of rain each day this week
sum(rainfall) / length(rainfall)  # Compute the mean the hard way
mean(rainfall)                    # Use a function to compute the mean
```

The first line creates a vector of seven values using the c() command, which either stands for "combine" or "concatenate," depending upon who you ask. These are the same rainfall numbers that I mentioned just above in the hand-calculated example. The second line calculates the mean using the definition I gave above, specifically to add up the total of all the data points with the sum() function and the length of the vector (the number of observations) with the length() function. The third line does the same thing, but uses a more convenient function, mean(), to do the job. The first line should produce no output, while the second two lines should each produce the value 1.142857. This is a great place to pause and make sure you understood the R commands we just used. These commands are the first time that you have used R to do some real work with data—an analysis of the central tendency of a week of rainfall data.

The Median

The **median** gets a fair amount of use, though not as much as the mean. The median is the "halfway value." If you sorted all of the elements in the vector from smallest to largest and then picked the one that is right in the middle, you would have the median. There is a function in R called median(), and you should try it on the rainfall vector, but before you do, can you guess what the result will be?

If you guessed "1," you are correct. If you sort the values in rainfall from left to right you get three 0's followed by two 1's, then the 2, then the 4. Right in the middle of that group of seven there is a "1," so that is the median. Note that the median of 1 is slightly smaller than the mean of about 1.14. This result highlights one of the important benefits of the median: it is resistant to **outliers.** An *outlier* is a case that someone who looks at the data considers to be sufficiently extreme (either very high or very low) that it is unusual. In the case of our 7 days of rainfall, the last day, with 4 centimeters of rain, was very unusual in comparison to the other days. The mean is somewhat enlarged by the presence of that one case, whereas the median represents a more typical middle ground.

The Mode

The **mode** is the value in the data that occurs most frequently. In some situations, the mode can be the best way of finding the "most typical" value in a vector of numbers because it picks out the value that occurs more often than any other. Like the median, the mode is very resistant to outliers. If you hear someone referring to "the modal value," he or she is talking about the mode and he or she means the value that occurs most frequently.

For odd historical reasons, the base installation of R does not contain a function that calculates the statistical mode. In fact, if you run mode(rainfall) instead of getting the statistical mode, you will learn some diagnostic information from R telling you what kind of data is stored in the vector called "rainfall." As always, though, someone has written an R package to solve this problem, so the following code will let you calculate the statistical mode on the rainfall vector. If you successfully followed the instructions in the Introduction, you may have already done these first two steps. Of course, there is no harm in running them again:

```
install.packages("modeest")    # Download the mode estimation package
library(modeest)               # Make the package ready to use
mfv(rainfall)                  # mfv stands for most frequent value
```

You will find that mfv(rainfall) reports 0 as the most frequent value, or what we refer to as the "statistical mode."

You might be wondering now how we are supposed to reconcile the fact that when we wanted to explore the central tendency of our vector of rainfall data we came up with three different answers. Although it may seem strange, having the three different answers actually tells us lots of useful things about our data. The mean provides a precise, mathematical view of the arithmetic center point of the data, but it can be overly influenced by the presence of one or more large or small values. The median is like putting the data on a see-saw: we learn which observation lies right at the balancing point. When we compare the mean and the median and find that they are different, we know that outliers are dragging the mean toward them. Finally, the mode shows us the particular value that occurs more frequently than any other value: 3 out of 7 days had zero rain. Now we know what is most common—though not necessarily most influential—in this sample of rainfall data.

In fact, the mode of 0 seems out of step with the mean rainfall of more than 1 centimeter per day, until we recognize that the 3 days without rain were offset by 4 other days with varying amounts of rain. With only seven values in our data set, we can of course eyeball the fact that there are 3 days with no rainfall and 4 days with rainfall, but as we start to work with bigger data sets, getting the different views on central tendency that mean, median, and mode provide will become very helpful.

MEASURES OF DISPERSION

I've noticed that a lot of people skip the introduction of a book. I hope you are not one of them! Just a few pages ago, in the Introduction, I mentioned a basic insight from the early days of applied statistics: that the central tendency of a set of data is important, but it only tells part of the story. Another part of the story arises from how dispersed, or "spread out," the data are. Here's a simple example. Three people are running for city council, but only the two candidates with the highest number of votes will get elected. In scenario one, candidate A gets 200 votes, candidate B gets 300 votes, and candidate C gets 400 votes. You can calculate in your head that the mean number of votes per candidate is 300. With these votes, most people would feel pretty comfortable with B and C taking their seats on the city council as they are clearly each more popular than A.

Now think about a second scenario: candidate X gets 299 votes, Y gets 300 votes, and Z gets 301 votes. In this case, the mean is still 300 and Y and Z would get elected, but if you think about the situation carefully you might be worried that neither Y nor Z is necessarily the people's favorite. All three candidates are almost equally popular. A very small miscount at the polls might easily have led to this particular result.

The mean of 300 is exactly the same in both scenarios—the only difference between scenario one and scenario two is the **dispersion** of the data. Scenario one has high dispersion and scenario two has low dispersion. We have several different ways of calculating measures of dispersion. The most commonly used measure of dispersion, the standard deviation, is the very last one mentioned in the material below because the others help to build a conceptual path on the way toward understanding the standard deviation.

The Range

The **range** is very simple to understand but gives us only a primitive view of dispersion. You can calculate the range simply by subtracting the smallest value in the data from the largest value. For example, in scenario one just above, 400–200 gives a range of 200 votes. In contrast, in scenario two, 301–299 gives a range of two votes. Scenario one has a substantially higher range than scenario two, and thus those data are more dispersed. You cannot rely on the range to give you a clear picture of dispersion, however, because one isolated value on either the high end or the low end can completely throw things off. Here's an example: There are 99 bottles of beer on the wall and 98 of them contain exactly 12 ounces of beer; the other bottle contains 64 ounces of beer. So the range is 52 ounces (64–12 = 52), but this entirely misrepresents the fact that the data set as a whole has virtually no variation.

Deviations from the Mean

You could think of the range in a slightly different way by considering how far each of the data points are from the mean rather than from each other. For scenario one, candidate A received 200 votes, and that is 100 votes *below* the mean. Candidate B received 300 votes and that is zero votes away from the mean. Finally, candidate C received 400 votes, and that is 100 votes *above* the mean. Instead of thinking of these three data points as having a range of 200 votes, you could say instead that there are two points at which each had a deviation from the mean of 100 votes. There was also one point that had 0 deviation from the mean.

Of course, it will get a bit unwieldy to describe these deviations if we have more than three data points to deal with. So what if we just added up all of the deviations from the mean? Sticking with scenario one, we add up these three deviations from the mean: −100 for candidate A plus 0 for candidate B plus +100 for candidate C. Oops, that adds up to 0! It will always add up to 0 for any collection of numbers because of the way that we calculated the mean in the first place. We could instead add together the **absolute values** of the deviations (i.e., make all of the negative values positive by removing the minus sign). Although this would get us going in the right direction, statistics researchers tried this a long time ago and it turns out to have some unhelpful mathematical properties (Cadwell, 1954). So we will get rid of the minus signs in another way, by **squaring** the deviations.

Sum of Squares

Still thinking about scenario one: candidate A is −100 from the mean; square that (multiply it by itself) to get 10,000. Candidate B is right on the mean, and zero times zero is zero. Finally, candidate C is +100 from the mean; square that to get 10,000. Now we can sum those three values: 10,000 + 0 + 10,000 equals 20,000. That quantity is the sum of the squared deviations from the mean, also commonly called the "**sum of squares.**" Sum of squares pops up all over the place in statistics. You can think of it as the totality of how far points are from the mean, with lots more oomph given to the points that are further away. To drive this point home, let's return to scenario two: candidate X is −1 from the mean; square that to get 1. Candidate Y is right smack on the mean; zero times zero is still zero. Finally, candidate Z is +1 from the mean; square that to get 1. Sum that: 1 + 0 + 1 equals 2. So the sum of squares for scenario one is 20,000 while the sum of squares for scenario two is 2: a humongous difference!

Variance

The sum of squares is great fun, but it has a crazy problem: it keeps growing bigger and bigger as you add more and more points to your data set. We had two scenarios with three candidates each, but what if one of our scenarios had four candidates instead? It wouldn't be a fair fight because the four-candidate data set would have one more squared deviation to throw into the sum of squares. We can solve this problem very easily by dividing the sum of squares by the number of observations. That is the definition of the **variance,** the sum of squared deviations from the mean divided by the number of observations. In a way, you can think of the variance as an average or "mean" squared deviation from the mean. Try this code in R

```
votes <- c(200,300,400)              # Here is scenario one again
(votes – mean(votes)) ^ 2            # Show a list of squared deviations
sum( (votes – mean(votes)) ^ 2)      # Add them together
# Divide by the number of observations
sum( (votes – mean(votes)) ^ 2) / length(votes)
```

Look over that code, run each line, and make sense of the output. If you are new to coding, the second line might blow your mind a bit because it shows how R can use one command to calculate a whole list of results. The little up-pointing character, ^, is commonly used in programming languages to raise one number to the power of another number. In this case ^ 2 squares the deviations by raising each deviation to the power of two. The final line calculates the variance: the result should be about 6666.667. You will sometimes hear researchers refer to the variance as the **mean square.** This makes sense because the variance is the mean of the squared deviations across the set of observations.

Note that I have presented four lines of code here with each one getting slightly more complicated than the previous one: The last line actually does all the work that we need and does not depend on the previous two lines. Those lines are there just to explain gradually what the code is doing. In most of our future R code examples, I won't bother with showing all of those intermediate steps in separate lines of code. Note, however, that this gradual method of coding is very advantageous and is used by experts and novices alike: begin with a very simple line of code, test it out, make it work, then add a little more complexity before testing again.

While we're on the subject, how about a little chat on the power of parentheses? Both mathematicians and computer programmers use parentheses to group items together. If you think carefully about the phrase "10 plus 10 divided by 5" you might realize that there are two different results depending upon how you interpret it. One way adds 10 and 10 and then divides that sum by 5 (making 4), while the other way takes "10 divided by 5" and then adds 10 to that result (making 12). In both math and programming it is easy to make a mistake when you expect one thing and the computer does another. So parentheses are important for avoiding those problems: (10 + 10)/5 makes everything nice and clear. As another example, in the second command shown in the block on page 12, (votes - mean(votes)) ^ 2, we gather (votes - mean(votes)) in parentheses to make sure that the subtraction is done first and the squaring is done last. When you type a left parenthesis on the command line in R, you will find that it helpfully adds the right parenthesis for you and keeps the typing cursor in between the two. R does this to help you keep each pair of parentheses matched.

Standard Deviation

The variance provides almost everything we need in a measure of dispersion, but it is a bit unwieldy because it appears in squared units. In our voting scenario, the value of 6666.667 is calibrated in "squared votes." While mathematicians may be comfortable talking about squared votes, most of the rest of us would like to have the result in just plain votes. Hey presto, we can just take the square root of the variance to get there. Try running this R code:

```
votes1 <- c(200,300,400)       # Here is scenario one again
# Here is the standard deviation
sqrt( sum((votes1 - mean(votes1))^2) / length(votes1) )
votes2 <- c(299,300,301)       # Here is scenario two
# And the same for scenario 2
sqrt( sum((votes2 - mean(votes2))^2) / length(votes2) )
```

Compare this to the previous block of code. The only difference is that we have enclosed the expression that calculates the variance within the sqrt() command. The square root of the variance is the **standard deviation,** and it is calibrated in the same units as the original data. In the case of scenario one our standard deviation is 81.65 votes, and for scenario two it is 0.8165 votes. This should feel pretty intuitive, as the scenario one deviations are a hundred times as big as the deviations in scenario two.

The standard deviation is a total workhorse among descriptive statistics: it is used over and over again in a wide range of situations and is itself the ingredient in many other statistical calculations. You will note that the R commands shown above are "chunky" and "clunky." R has a built-in var() function for variance and a built-in sd() function for standard deviation that you can use from now on. For reasons that we will encounter later in the book, these functions will produce variances and standard deviations that are slightly larger than the ones produced by the calculation shown on page 13. The calculation shown in the R code above is the so-called "population" standard deviation, whereas sd() provides the "sample" standard deviation. In large data sets these differences are negligible. If you want to experiment with var() and sd() in R, your conceptual understanding of what they signify is perfectly valid even though the calculations used inside of the var() and sd() functions are ever so slightly different than what we used in the material discussed earlier.

Mean and Standard Deviation Formulas

Although this book explains statistical ideas without relying on formulas, having some knowledge of formulas and the ability to read them may be helpful to you later on. For that reason, I include a few of the most common formulas so that you can practice interpreting them and so that you can recognize them later in other contexts. Let's get started with two of the easiest formulas, the population mean and the population standard deviation:

Mean:
$$\mu = \frac{\sum x}{N}$$

The equation for the mean says that mu equals the sum of all x (all of the data points) divided by N (the number of data points in the population). The small Greek letter mu looks like a u with a little downward tail on the left and stands for the population mean. The capital Greek letter sigma on the right-hand side of the equation looks a bit like a capital E and here signifies that all of the x's should be added together. For this reason, the capital sigma is also called the summation symbol.

Standard deviation:
$$\sigma = \sqrt{\frac{\sum (x-\mu)^2}{N}}$$

The equation for the standard deviation says that sigma (the population standard deviation) equals the square root of the following expression: subtract the population mean (mu) from each x and square the result. Calculate a sum of all these squared deviations and divide by N, the number of data points in the population. Here we learned a new Greek letter, the small sigma, which looks like an o with a sideways tail on top pointing right. Make a note to yourself that the small letter sigma indicates the standard deviation and the capital letter sigma is the summation symbol.

DISTRIBUTIONS AND THEIR SHAPES

Measures of central tendency and of dispersion, most commonly the mean and the standard deviation, can be applied to a wide variety of data sets, but not to all data sets. Whether or not you can apply them depends a lot on the **distribution** of the data. You can describe a distribution in words—it is the amount of observations at different spots along a range or continuum, but that really does not do justice to the concept. The easiest way for most people to think about distributions is as shapes. Let's explore this with two different types of distributions, the normal distribution and the Poisson (pronounced "pwa-sonn") distribution.

The Normal Distribution

Many people have heard of the **normal distribution,** although it is commonly called "the bell curve" because its shape looks like the profile of a bell. The common explanation for why a distribution of any measurement takes on a bell shape is that the underlying phenomenon has many small influences that add up to a combined result that has many cases that fall near the mean. For example, heights of individual people are normally distributed because there are a variety of different genes controlling bone shape and length in the neck, torso, hips, thighs, calves, and so forth. Variations in these lengths tend to cancel each other out, such that a person of average height has data that falls in the middle of the distribution. In a few cases many or all of the bones are a little shorter, making a person who is shorter, and therefore in the lower part of the distribution. In some of the other cases, many or all of the bones are a little longer, making a person who is taller and therefore in the higher part of the distribution. These varying shapes and lengths add together to form the overall height of the person. Try this line of R code:

```
hist( rnorm(n=1000, mean=100, sd=10) )
```

This line runs the rnorm() function, which generates a list of 1,000 random numbers in a normal distribution with a specified mean (in this case 100) and a standard deviation (in this case 10). The hist() function then takes that list of points and creates a histogram from them—your first graphical output from R! You should get a plot that looks something like Figure 1.1. Your plot may look a little different because the random numbers that R spits out are a little different each time you run the rnorm() command.

The term "histogram" was supposedly coined by Karl Pearson, another 19th-century statistician (Magnello, 2009). The histogram comprises a series of bars along a number line (on the horizontal or X-axis), where the height of each bar indicates that there are a certain number of observations, as shown on the Y-axis (the vertical axis). This histogram is considered a **univariate** display, because in its typical form it shows the shape of the distribution for just a single variable. The points are organized in groupings from the smallest values (such

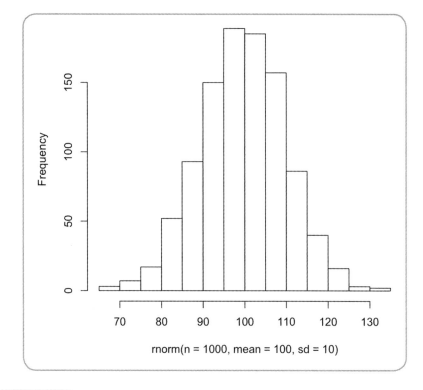

FIGURE 1.1. Histogram of *n* = 1,000 random points in a normal distribution with a mean of 100 and a standard deviation of 10.

as those near 70 in the graph in Figure 1.1) to the largest values (near 130 in the graph in Figure 1.1). Each of the groupings must have the same range so that each bar in the histogram covers the same width (e.g., about 5 points in Figure 1.1). These rising values for each subsequent group go from left to right on the *X*-axis (horizontal) of the histogram. If you drew a smooth line connecting the top of each bar in the histogram, you would have a curve that looks a bit like the outline of a bell. The two or three bars above the low end of the *X*-axis are the left-hand **tail,** while the top two or three bars on the high side are the right-hand tail. Later in the book we will have more precise boundaries for what we include in the tails versus what is in the middle.

The histogram in Figure 1.1 includes one bar for each of about 14 discrete categories of data. For instance, one category might include a count of all of the data points between 100 and 105. Because we only have 14 categories, the shape of the profile across the tops of the bars is blocky, as if you were looking at a set of stairs. Now imagine if instead of just 14 categories we had 100 categories. You can try this yourself in R with this command:

```
hist( rnorm(n=10000, mean=100, sd=10), breaks=100 )
```

With 100 categories, the profile of the histogram begins to look much smoother because the bars are thinner and there's a lot more of them. If you kept going with this process to make 1,000 categories/bars, the top of the histogram would start to look like a smooth curve. With an infinite number of categories and an infinite amount of normally distributed data, the curve would be perfectly smooth. That would be the "ideal" normal curve and it is that ideal that is used as a model for phenomena that we believe are normally distributed.

For example, the histogram in Figure 1.1 closely matches the results that come from many tests of general intelligence. Many of these tests are scored in a way such that a typical person gets a score near 100, and a person who is one standard deviation higher than the average person has a score of 110. Large studies of general intelligence often show this bell-shaped distribution of scores, so the normal curve provides a model for the scores obtained from this kind of test. Note that my use of the word *model* is very important here. If you think back to the Introduction, you may remember that my definition of statistical significance referred to thinking about chance occurrences and likelihoods in terms of a model. The normal distribution is one of these models—in fact probably the single most popular model for reasoning about statistical significance.

The Poisson Distribution

Not so many people have heard of the Poisson distribution, named after the 19th-century French mathematician Simeon Poisson. Like the normal distribution, the Poisson distribution fits a range of natural phenomena, and in particular works well for modeling arrival times. A famous study by the 19th-century Russian economist Ladislaus von Bortkiewicz examined data on military horsemen who died after being kicked by their horses (Hertz, 2001). Ouch! Von Bortkiewicz studied each of 14 cavalry corps over a period of 20 years, noting when horsemen died each year. Imagine a timeline that starts on January 1, 1850. If the first kick-death was on January 3, and the second was on January 4 we would have two data points: the first death occurred after 2 days and the second death occurred after one additional day. Keep going, and compile a whole list of the number of days between kick-death events. The distribution of these "arrival" times turns out to have many similarities to other kinds of arrival time data, such as the arrival of buses or subway cars at a station, the arrival of customers at a cash register, or the occurrence of telephone calls at a particular exchange. The reason that these phenomena share this similarity is that in each case a mixture of random influences impacts when an event occurs. You can generate a histogram of some random arrival data using the following line of R code:

```
hist( rpois(n=1000, lambda=1) )
```

Note that the rpois() function, which generates a list of random numbers that closely match a Poisson distribution, has two parameters. The first parameter, n, tells the function how many data points to generate. Note in this case that we have asked for 1,000 data points. The lambda parameter is the value of the mean—lambda is a Greek letter that statisticians like to use. The idea is that if you calculated a mean of the 1,000 data points it should come out to roughly 1. Note that it will almost never come out exactly to 1 because the data points are random and we have only generated a finite number of them. If we had an infinitely powerful computer, we might calculate the mean on an infinite set of numbers and it would come out to exactly one. So generally, you will expect a close match to the target value that you use for lambda but not an exact match. Finally, note that the code above has the rpois() function nested inside the hist() function in order to immediately create a histogram. The result should look something like Figure 1.2.

How should one interpret the histogram in Figure 1.2? Remember that one way of thinking about values that make a Poisson distribution is thinking of them as delays between arrival times—in other words the interval of time between two neighboring arrivals. So if we imagined 1,000 cars going past a particular location on an active road, a lot of them go by with 0 minutes, 1 minute, or 2 minutes between them, with the most common (modal) case being 0 minutes between cars (about 360 out of our 1,000 cases). A turtle who took 4 or 5 minutes to cross the road would be in big trouble: only a few of the cases show an interval of 4 minutes (roughly 20 cases per 1,000) or 5 or 6 minutes (just one or two cases per 1,000). If you are a turtle crossing this particular road, you should probably invest in a jet pack.

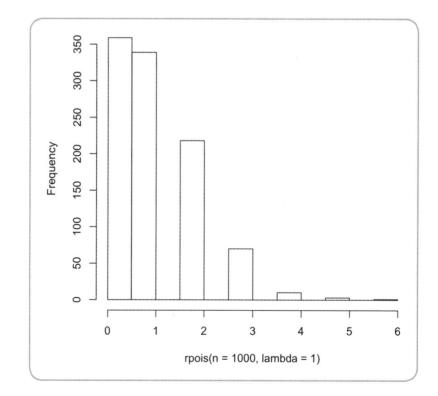

FIGURE 1.2. Histogram of n = 1,000 random points in a Poisson distribution with a mean of 1.

By the way, you can check for yourself that the mean of the distribution comes out to about 1. Instead of embedding the rpois() command in hist(), embed it in the mean() function, like this:

```
mean(rpois(n=1000, lambda=1))
```

One time when I ran this command, I got a mean of 1.032. Another time I got 0.962. This little bit of variation is the natural result of using random numbers and you should not be worried that there is anything wrong with your computer or code. What command could you use to find out the standard deviation of your Poisson distribution? (Hint: look back to the earlier part of this chapter when we discussed measures of dispersion.) Try it and see what you get. Then look up some more information about the Poisson distribution to see if you can explain your results.

If you find it annoying to have different results every time you run rpois() (or any other random number generator) you could generate a distribution just once and stash it away in a vector for reuse in later commands:

```
myPoiSample <- rpois(n=1000, lambda=1)
mean(myPoiSample)
hist(myPoiSample)
```

Remember that these two characters together

```
<-
```

are the so-called "assignment arrow" that is commonly used in R. I like the assignment arrow because it makes it quite clear that the stuff that is calculated on the right-hand side is going into the thing on the left-hand side. By the way, it is common practice in almost all programming languages to have the expression on the right assigned to the thing on the left, although most languages use an equal sign instead of the assignment arrow. The equal sign also works in R if you prefer it.

CONCLUSION

In this chapter, I introduced some essential statistical terms and concepts including the mean, the median, the mode, the range, variance, the standard deviation, and two types of distributions: the normal distribution and the Poisson distribution. All of these terms help us to think about and summarize individual vectors (lists) of numbers—what statisticians and computer scientists call **variables.** For this reason, much of what I have discussed here is known as **univariate descriptive statistics.** In other words, we are using statistics to describe single variables. Single variables are the main building blocks of larger statistical data sets, and while you can learn a lot from just looking at one variable at a time, most of the fun of playing with data comes from look-ing at two or more variables together. In the next chapter, we look at some of the ideas behind probability and, as you will see, probability starts us into the process of thinking about two variables together.

EXERCISES

1. Using the material from this chapter and possibly other information that you look up, write a brief definition of these terms in your own words: **mean, median, mode, variance, standard deviation, histogram, normal distribution,** and **Poisson distribution.**

2. Write the equations, using the appropriate Greek letters, for the population mean and population standard deviation. Explain briefly what each Greek letter means. The R environment offers about 20 different kinds of statistical distributions. Choose any one of these distributions other than the normal distribution or the Poisson distribution. (The help system in R can assist you with finding a description of these distributions and their commands: type "?distributions" at the command line. For a hint about one distribution you might choose to study, read the beginning of the next chapter!) Write some R code that generates 100 random points in that distribution, displays a histogram of those 100 points, calculates the mean of those points, and calculates the standard deviation. Make sure to use the technique shown just above that begins with assigning the 100 points to a vector that can be reused for all of the other commands.

3. Use the data() function to get a list of the data sets that are included with the basic installation of R: just type "data()" at the command line and press enter. Choose a data set from the list that contains at least one numeric variable—for example, the Bio-chemical Oxygen Demand (BOD) data set. Use the summary() command to summarize the variables in the data set you selected—for example, summary(BOD). Write a brief description of the mean and median of each numeric variable in the data set. Make sure you define what a "mean" and a "median" are, that is, the technical definition and practical meaning of each of these quantities.

4. As in the previous exercise, use the data() function to get a list of the data sets that are included with the basic installation of R. Choose a data set that includes just one variable, for example, the LakeHuron data set (levels of Lake Huron in the years 1875 through 1972). Use the hist() command to create a histogram of the variable—for example, hist(LakeHuron). Describe the shape of the histogram in words. Which of the distribution types do you think these data fit most closely (e.g., normal, Poisson). Speculate on why your selected data may fit that distribution.

CHAPTER 2

Reasoning with Probability

Probability is a branch of mathematics that scares the you-know-what out of many people and yet everyone who has ever played a game of cards or rolled a pair of dice has used probability in an easy and intuitive way. If you think about why cards and dice are so popular, it is partly because the human brain loves to think about probabilities intuitively and make guesses and decisions about chance events. Going way, way back into prehistory, we might have had to guess at the likelihood of whether a sleeping lion would wake up and chase us or if those dark clouds signaled a rain storm. The sleeping lion example illustrates why probability is so important to us: the world is full of uncertainty, and reasoning about probability gives us an important tool to understand and manage that uncertainty. Over the past few centuries mathematicians have applied their craft to understanding probability and to developing tools that let us understand with precision how randomness and probability connect with each other. They have developed derivations, proofs, and equations for all of the important principles involved in reasoning about chance events, but we will not focus on those in this book. Instead, we will use tables and graphs to reason about probability using some straightforward examples that most people can relate to quite easily.

OUTCOME TABLES

The simplest and easiest-to-use tool for thinking about probability is an **outcome table.** An outcome table lists the various possible outcomes of a set of similar or related events. The most common example used in most textbooks is a coin toss, but let's change things up a bit and instead think about a "toast drop." Imagine a piece of toast with jelly on it. Mmmm, jelly! The toast falls from a dish and lands on the kitchen floor. This dropping event has just two possible outcomes: jelly-side-up (yay) and jelly-side-down (boo). When there are two possible outcomes of an event, statisticians refer to it as **binomial,** meaning "two names."

Now, if you just drop the toast one time, it is difficult to predict what will happen, but what if you dropped the toast six times? We could call this a **trial**—in this example the trial consists of six events, each of which has two possible outcomes. What could happen in one trial of six events? One possibility would be that the toast landed with the jelly-side down three times and with the jelly-side up three times. But that is not the only configuration, of course. I think it might not be unusual to get jelly-side-down four times and jelly-side-up twice, or perhaps the other way around. One thing that seems highly unlikely is to get jelly-side-up six times in a row. Likewise, it would also be really unusual (and quite messy) to have the jelly hit the floor six times in a row.

As the description above suggests, some events are more likely to occur than others, and the **binomial distribution** models these possibilities across multiple trials. We can use R to produce some random binomial trials for us:

```
table( rbinom(n=100,size=6,prob=0.5) )
```

The rbinom() function produces random outcomes in binomial distributions. The size=6 parameter tells the function that we want six toast-drop events in each trial. Remember that we are using the word *trial* to refer to a set of tests, experiments, or events that is always considered as a group. The *n*=100 parameter tells the function to run 100 trials. In this case the prob=0.5 indicates that we believe our toast is equally likely to fall with the jelly down or up in any single event. There are lots of silly jokes about it being more likely for the jelly side to fall down, so later on you might want to experiment with changing prob=0.5 to some other value. The table() function that is wrapped around the outside neatly summarizes the resulting list of trials. I ran this command and got the results that appear in Table 2.1.

The top line of Table 2.1 shows the number of times that the jelly side landed on the floor. For any one trial of six toast drops we do not need to show the number of times that the jelly landed up because it is always six minus the number of jelly-downs.

You might find it surprising to realize that there are actually seven possible scenarios! Why? As a hint, think carefully about the meaning of the zero column that is the leftmost category in Table 2.1.

TABLE 2.1. 100 Toast-Drop Trials, 6 Drops per Trial							
Jelly-down count	0	1	2	3	4	5	6
Number of trials with that count	4	9	20	34	21	11	1

The second line summarizes what happened: how many trials came out that way. Out of 100 trials, there were only four trials where there were no jelly-down events. Make sure you can read and make sense out of this table. How many trials had three jelly-down events?

It is easy to figure out the probability of each category of event. I hope you are as glad as I am that we ran 100 trials: it makes the calculations easy because you can just divide each event count by 100 to get the probability of each kind of event shown in the bottom row of Table 2.2.

Much as we expected, three jelly-down events is the most likely outcome: it alone accounts for more than a third of all trials (0.34, or 34% of all trials). Note that its neighbors, two jelly-downs and four jelly-downs, are also quite likely. I think in some ways it is easier to look at a graph that represents this kind of data. Use the following command to generate a histogram of toast-drop outcomes:

```
hist( rbinom(n=100,size=6,prob=0.5) )
```

You should remember that your results will look slightly different from the histogram shown in Figure 2.1 because rbinom() generates random numbers such that the trial outcomes will generally be different each time. Occasionally you may not even get anything in the zero category or the six category. For this reason you might want to run 1,000 trials (or more) to get a better looking graph. Here is the command (look in Figure 2.1 for the results):

```
hist( rbinom(n=1000,size=6,prob=0.5) )
```

What an interesting graph! It should remind you of something from earlier in the book. There is something a little odd, however, about how R shows the 0 and 1 categories right next to each other with spaces between the other categories, and that has to do with the fact that there are no intermediate values (like 1.5). We can get a much nicer display by using the barplot() function to plot the output of the table command, as shown in Figure 2.2:

```
barplot( table( rbinom(n=1000,size=6,prob=0.5) ) )
```

The barplot() function simply requires a list of individual values to tell R how high each bar should be. For this job that simplicity helps us. In fact, using the barplot() function and the table() function together gives us a neat

TABLE 2.2. 100 Toast-Drop Trials, 6 Drops per Trial, with Probabilities							
Jelly-down count	0	1	2	3	4	5	6
Number of trials with that count	4	9	20	34	21	11	1
Probability of that count	.04	.09	.2	.34	.21	.11	.01

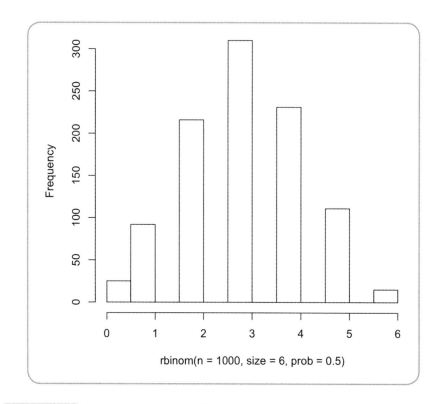

FIGURE 2.1. Histogram of $n = 1,000$ random trials in a binomial distribution.

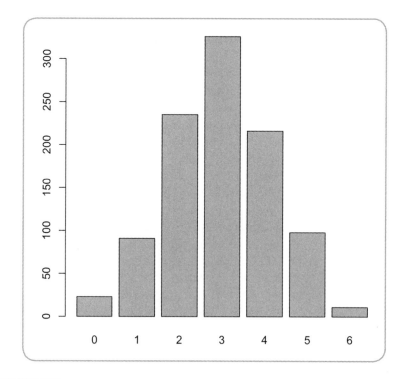

FIGURE 2.2. Bar plot of $n = 1,000$ random trials in a binomial distribution.

new option, as shown in Figure 2.3. If we divide each value in the table by the total number of trials, we can use the bar plot to display probabilities instead of counts. Try the table() by itself and then use the barplot() function to show it graphically, using these two commands:

```
table(rbinom(n=1000,size=6,prob=0.5))/1000
barplot(table(rbinom(n=1000,size=6,prob=0.5))/1000)
```

You will notice in Figure 2.3 that the only thing that changed (other than a slight variation in bar heights, because it was a new set of random trials) is the Y-axis. The Y-axis now shows probabilities instead of raw counts. We can read from this graph, for instance, that the probability of five jelly-down events is about 0.10 or 10%. You can also combine events. For example, the probability of getting four or more jelly-down events seems to be about 35% (add up the bars for 4, 5, and 6). Of course, if we wanted to be really exact about combining the probabilities of events, we could go back to using a table instead of a graph. We can also get R to do the work for us.

```
probTable <- table(rbinom(n=1000,size=6,prob=0.5))/1000
probTable
cumsum(probTable)
```

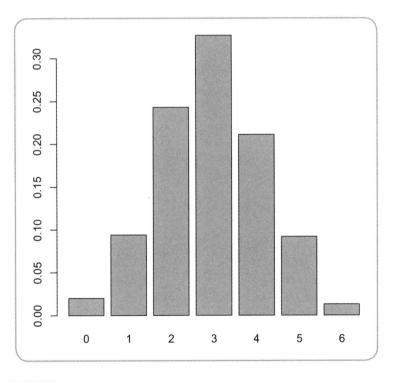

FIGURE 2.3. Bar plot of probabilities of each trial for $n = 1,000$ random trials in a binomial distribution.

First we create a table that summarizes the probabilities for 1,000 trials of six toast drops each. The second line simply reports out what the variable probTable contains. The third line uses a "cumulative sum" function to add each successive probability to the sum of the previous ones. To explain a cumulative sum with a simple example, let's say you had a list of three numbers: 1, 3, and 5. The three cumulative sums from this list are 1, (1+3), and (1+3+5), in other words 1, 4, and 9. For any given set of event probabilities, the cumulative probabilities should always add to 1. Look at Table 2.3 to see what I got when I ran that code.

Now you may think me unusual, but I also think that it will be fun to make a bar plot of the cumulative sum of the probabilities:

```
probTable <- table( rbinom(n=1000,size=6,prob=0.5) )/1000
barplot( cumsum(probTable) )
```

That code yields the bar plot shown in Figure 2.4.

So what's the big deal about the bar plot in Figure 2.4, you might ask. The cool thing about this plot (and/or its corresponding table of cumulative probabilities) is it allows us to reason easily about a *range* of outcomes rather than just a single outcome. For example, what is the probability of having two *or fewer* jelly-down outcomes? You can see from the graph that the probability is about 0.33 or 33%. Another way of saying the same thing is to say that if you add up the probability of getting zero jelly-downs, one jelly-down, and two jelly-downs you will account for about one-third of all of the possible outcomes. Here's an easy question that you should be able to answer with your eyes closed: what's the probability of six or fewer jelly-down events? That's easy, because six or fewer jelly-downs accounts for all possible outcomes, so the cumulative probability must be one. Before moving on, make sure you have taken a close look at that bar plot and that you can reason about any of the categories shown on the X-axis of the plot. Cumulative probabilities will come in really handy later on when we are thinking about the area under a distribution curve.

There's one last thing that you should try on your own before we move on to another topic. For each of the examples above, we considered one trial to consist of six toast-drop events (controlled by the parameter size=6 in the rbinom() command above). I chose six events because it makes it more convenient to talk about a limited range of jelly-down and jelly-up outcomes for each trial. We were able to look at the whole set in one brief table. Consider what would happen, however, if each trial consisted of 100 toast drops or even 1,000 toast drops. What do you think the bar plots would look like then? By this

TABLE 2.3. 100 Toast-Drop Trials, 6 Drops per Trial, with Cumulative Probabilities							
Jelly-down count	0	1	2	3	4	5	6
Number of trials with that count	.015	.101	.33	.64	.903	.985	1

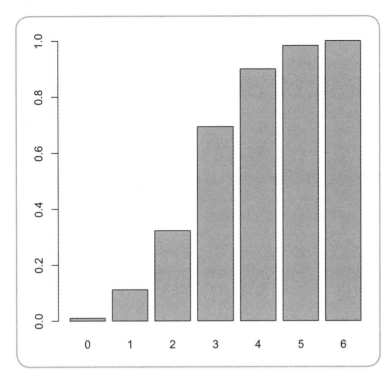

FIGURE 2.4. Bar plot of cumulative probabilities for n = 1,000 random trials in a binomial distribution.

point in the chapter, you should be able to formulate your own R commands to try these ideas out, but here's some code to get you started:

```
probTable <- table(rbinom(n=10000,size=100,prob=0.5))/10000
barplot(probTable)
barplot(cumsum(probTable))
```

You should be able to read this code and see that I have now set up each trial to consist of 100 toast drops and I have asked R to run 10,000 trials. You should be able to modify this code both to produce a different number of toast drops per trial as well as a different number of trials. Also, you should be able to make sense out of the resulting bar plots of probabilities and cumulative probabilities. For each of these two bar plots you should be able to point to a spot on the X-axis and say what the height of the bar at that point in the plot signifies. Make sure this skill is solid in your mind as we will use it extensively later in the book.

CONTINGENCY TABLES

Besides the simple outcome table, another of the important—and easy to use— tools for thinking about probability is the contingency table. A **contingency**

table is a tool for asking and answering "what if" questions about a more complex, linked set of outcomes than what we were considering before. Contingency tables must have at least four cells. For the most part these contingency tables are rectangular and they have rows and columns. A simple four-cell contingency table with some helpful labels appears in Table 2.4.

Pretty easy stuff: columns go across the top, rows go down the side, and each cell lies at the intersection of exactly one row and exactly one column. We can take the same table and label it with some real names so that we can do some examples as in Table 2.5.

In Table 2.5, we are extending our previous toast-drop example by thinking about a very clumsy waiter who drops different types of toast on the kitchen floor. Here we have added a new layer of complexity over the simple jelly-up/jelly-down outcomes we were considering earlier in the chapter. Some toast now has butter on it (the first row of the table) while other toast has jelly on it (the second row of the table). As the clumsy waiter drops the toast, it either lands on the floor with the topping facing down (e.g., the jelly side down) or with the topping facing up (e.g., the butter on top). When we put some data into this table to summarize all of the dropped toast from this morning's orders, we can refer to each cell individually. For example, cell 2B will contain a count of all of the times that the waiter dropped toast with jelly and it landed jelly-side-up.

In this particular example, we are not at all concerned with toast that was not dropped and we are not concerned with dry toast, so you should ignore those kinds of outcomes. If it helps, think of this as a reality TV comedy where the waiter *always* drops the toast and the toast *always* has a topping: either butter or jelly. For the sake of entertainment, we are completely ignoring competent waiters and nonmessy foods.

Table 2.6 has some data that summarizes the morning's clumsiness.

There are a lot of new and different questions we can answer using this very simple data display. Here are a few of the most important ones:

TABLE 2.4. Generic Two-by-Two Contingency Table		
	Column A	**Column B**
Row 1	Cell 1A	Cell 1B
Row 2	Cell 2A	Cell 2B

TABLE 2.5. Generic Two-by-Two Contingency Table with Toasty Labels		
	Down	**Up**
Butter	Cell 1A	Cell 1B
Jelly	Cell 2A	Cell 2B

TABLE 2.6. Two-by-Two Contingency Table with Toasty Labels and Data

	Down	Up
Jelly	2	1
Butter	3	4

- What was the total number of "toast-drop" events that occurred this morning?
- How many times did the toast land with the topping down?
- How many times did the toast land with the topping up?
- Was it more likely that the toast lands with the topping down or up?
- Across all of our dropped toast, which was more likely, butter or jelly?
- Did the topping make any difference in how the toast landed?

You can probably think of at least a couple more questions. Before we start to answer these questions, however, we can improve the usefulness of our table with a couple of easy modifications. First, I am going to add marginal row totals on the right and marginal column totals on the bottom, as shown in Table 2.7.

It should be pretty easy to see how this works in Table 2.7. At the bottom of each column, I have put in a sum that adds up the values in all of the rows for that one column. Likewise at the end of each row, I have added a sum across all of the columns. These are called "marginal totals" simply because they are in the "margins" (i.e., the outer edges) of the table. Finally, in the lower right-hand corner, there is a grand total representing all of the events. This grand total is very important for the next step.

We are going to replace each cell count with a corresponding probability. This follows the same procedure we used with the simple outcomes table: just divide each cell count by the total number of events. The result appears in Table 2.8.

Now we are equipped to answer a range of questions about probabilities. For example, what is the most likely event among the four basic outcomes? Ignoring the marginal totals for the moment, you should be able to see that butter-side-up is the single most likely combination: 40% of the toast-drop

TABLE 2.7. Two-by-Two Contingency Table with Toasty Labels, Data, and Marginal Totals

	Down	Up	Row totals
Jelly	2	1	3
Butter	3	4	7
Column totals	5	5	10

TABLE 2.8. Two-by-Two Contingency Table with Frequencies Converted to Probabilities			
	Down	**Up**	**Row totals**
Jelly	.2	.1	.3
Butter	.3	.4	.7
Column totals	.5	.5	1.0

events this morning resulted in butter-side-up. Of course, the marginal totals give us lots more to think about as well. For example, across all dropped toast which kind of topping seems to be more popular, butter or jelly? A look at the final column shows that there is a 70% chance of buttered toast and only a 30% chance of jelly toast among all of the toast that was dropped this morning. (I keep saying "this morning" to emphasize that these 10 toast-drop trials are just one such possible set; if we collected our clumsy waiter data again tomorrow, we might get a different bunch of results. Also don't forget that we are only

Make Your Own Tables with R

The tables in this chapter were so simple to create that I made them by hand in my word processor. You can use R to create your own tables, however, and even do all of the calculations of margin totals and probabilities automatically. Here's some code that re-creates most of the information shown in Tables 2.7 and 2.8:

```
# Create a two-column, two-row structure using the matrix() command.
# This line of code fills the cells row-by-row, so the first row gets 2,1
# and the second row gets 3,4.
toast <- matrix(c(2,1,3,4),ncol=2,byrow=TRUE)
colnames(toast) <- c("Down","Up")         # Label the columns
rownames(toast) <- c("Jelly","Butter")    # Label the rows
toast <- as.table(toast)                   # Convert from matrix to table
toast                                      # Show the table on the console
margin.table(toast)                # This is the grand total of toast drops
margin.table(toast,1)              # These are the marginal totals for rows
margin.table(toast,2)              # These are marginal totals for columns
toastProbs <- toast/margin.table(toast)   # Calculate probabilities
toastProbs                                 # Report probabilities to console
```

Remember that you do not have to type anything that comes after the # character: all of that just provides explanatory comments. When you get to the last line, you will find that toastProbs contains a two-by-two table of the probabilities just like Table 2.8, except for the marginal total probabilities. Given what you have learned from the previous lines of code, can you write three more lines of code that provide the grand total and marginal total probabilities?

thinking about the clumsy waiter: all toast orders are topped and dropped in our scenario.)

You should be able to look at the cell probabilities and marginal probabilities and answer a number of different questions. Is it more likely that toast will land with the topping down or the topping up? If you wanted to avoid a topping-down event, which topping should you choose?

Let's say you wanted to predict what would happen to the next order of dropped toast, *which you know has butter on it*. Which outcome is more likely, butter-down or butter-up? Similarly, let's say you notice that there is a piece of toast on the floor *with the topping side down* (so you can't see what the topping is). If you had to bet on jelly or butter, which would you choose?

These last two questions are important because they introduce a new principle that will come in handy later on: When we are presented with a new piece of information about a probability scenario, how does it change what we believe? In the previous paragraph, the italicized words represent new information we have about the situation. That new information helps us to get better at pinpointing our probabilities. In Table 2.8 if you know that the toast has butter on it, you can ignore the jelly row and just focus on the probabilities in the butter row.

Let's kick this idea up one more notch with a final example from our crazy toast restaurant. In the table below we have more cells, but the essential ideas are the same as for the jelly-butter/up-down example. Table 2.9 contains the work schedules of the four clumsy servers who work here and drop your toast.

By now you should be a total expert at making sense out of these tables. Each cell contains a number of work hours for a given person on a given day. There are also marginal column totals for each person and marginal row totals for each day. Finally, the lower right-hand corner contains the grand total of work hours for a whole week across all servers. What should we do next to make this table more useful?

Exactly right: we should *normalize* the table by dividing each cell by the grand total. This will turn each count of hours into a probability. In this particular case you might find it more natural to think of each cell as a *proportion* of

TABLE 2.9. Larger Contingency Table of Days and Server Hours				
Work hours	**Anne**	**Bill**	**Calvin**	**Devon**	**Marginal**
Monday	6	0	5	8	19
Tuesday	0	6	6	0	12
Wednesday	8	5	0	8	21
Thursday	4	6	5	8	23
Friday	8	8	5	8	29
Saturday	8	10	8	8	34
Sunday	0	3	6	0	9
Marginal	34	38	35	40	**147**

the total work hours, but mathematically that is the same as a probability. Table 2.10 shows what the table looks like after normalizing.

You can see that certain servers have certain days off, and these appear as zero probability cells. Note that I have chosen to use four significant digits in this display simply because some of the probabilities are fairly small. As a result there will be some rounding error if you try to calculate the marginal totals from the data shown here. When you are working with these data in R, the program will automatically represent each probability with high numeric precision.

Here is the new, important, and slightly tricky question. Actually there are two related questions, one of which involves choosing a column and one of which involves choosing a row:

1. Let's say you are feeling just a little bit dazed and confused and you have forgotten what day it is. You sit down in the toast restaurant and you notice that Bill is your server. What day is it most likely to be? Once you have answered that question, here's the tricky part: Given that Bill is your server, what is the specific probability that it is the day that you chose?

2. Today you are not confused at all—it is most definitely Monday. If you go to the restaurant today, knowing that it is Monday, who is most likely to be your server? What is the specific probability that your server will be that person?

To address these questions, let's isolate a single row or column and focus our attention only on that one piece of the table. In essence, when we are given a new piece of information, such as "Your server is Bill," this helps us focus our attention on just a subset of all of the possible outcomes. Knowing that our server is Bill allows us to address the question "What day is it?" with more certainty than we could have without this additional knowledge. Bill's column is separated out in Table 2.11.

TABLE 2.10. Larger Contingency Table with Server Hours Converted to Probabilities

Normalized	Anne	Bill	Calvin	Devon	Marginal
Monday	.0408	.0000	.0340	.0544	.1290
Tuesday	.0000	.0408	.0408	.0000	.0820
Wednesday	.0544	.0340	.0000	.0544	.1430
Thursday	.0272	.0408	.0340	.0544	.1560
Friday	.0544	.0544	.0340	.0544	.1970
Saturday	.0544	.0680	.0544	.0544	.2310
Sunday	.0000	.0204	.0408	.0000	.0610
Marginal	.2310	.2590	.2380	.2720	1

TABLE 2.11. Excerpt of Table Showing Server Bill Only	
Bill is our server: What day is it?	Bill's raw work probabilities
Monday	.0000
Tuesday	.0408
Wednesday	.0340
Thursday	.0408
Friday	.0544
Saturday	**.0680**
Sunday	.0204
Marginal	.2590

Even if I had not used bold type in Saturday's entry in Bill's column, you should be able to see very quickly that the entry for Saturday has the highest value among all 7 days. So if we did not know what day it was, and we did get Bill as our waiter, it should be obvious that it is most likely to be Saturday. But what is the specific probability that it is Saturday? Don't say 0.0680! That's incorrect. In fact, 0.0680 is the probability of having Bill as your server on Saturday *given no advance knowledge of what day it is or who your server will be.* If that sounds confusing, think of it this way: Imagine 1,000 raffle tickets in a bag, and you get to pick one ticket for a free slice of toast at the restaurant (naturally, your toast will be dropped on the floor before you get it). Each ticket is marked with a specific day of the week on which you must visit the restaurant, as well as the particular server who must bring out your free toast from the kitchen. The bag contains exactly 68 tickets that are marked Bill/Saturday, so your chances of picking one of those from among 1,000 total raffle tickets is 0.0680. By the way, I learned this method for explaining a probability problem using real things like lottery tickets from a great book entitled *Calculated Risks* (Gigerenzer, 2002). Check out a copy at your library!

Now back to our problem. Look at the bottom row of Table 2.10: Anne's shifts cover 23.1% of all work hours, Calvin's cover 23.8%, and Devon's cover 27.2%. Because we know that our server is Bill, the proportion of time attributable to Anne, Calvin, and Devon is now irrelevant and can be ignored. That means that the remaining proportion of work hours—the 25.9% attributable to Bill—now represents the totality of possibilities for us. So we need to scale up all of Bill's probabilities to make the original 25.9% now cover 100%. If you have your thinking cap on, you might see that the way to do this is to normalize Bill's Raw Work Probabilities column based on the marginal probability of our server being Bill (in other words, 0.2590). To get the normalized probabilities, all we have to do is divide each value in the Bill column by 0.2590. Table 2.12 adds a column to Bill's original table that contains the normalized probabilities.

TABLE 2.12. Excerpt of Table Showing Server Bill Only with Normalized Probabilities		
Bill is our server: What day is it?	Bill's raw work probabilities	Bill's normalized work probabilities
Monday	.0000	.0000
Tuesday	.0408	.1579
Wednesday	.0340	.1316
Thursday	.0408	.1579
Friday	.0544	.2105
Saturday	.0680	.2632
Sunday	.0204	.0789
Marginal	.2590	1

Now we are equipped to answer the question: given the advance knowledge that our server is Bill, the probability that it is Saturday is 0.2632, or a little above 26%. Of course we can also see the probabilities of all of the other days. Given that our server is Bill, we can be certain that it is not Monday, right? By the way, look again in the bottom right cell—it contains the normalized marginal probability across all days of the week. Intuitively it should make sense that the cell contains 1, but can you explain in your own words what that means? One way to phrase it is that it is the overall probability of having Bill as your waiter across all days of the week, given advance knowledge that your waiter is Bill. While technically true, it sounds kind of silly. Instead, you might better phrase it as the last entry in a list of *cumulative* probabilities as the week moves onward through all 7 days. In fact if you harken back to our earlier discussion of outcome tables, 1.0 is the cumulative probability that it is *Sunday or earlier in the week*, given the knowledge that Bill is our waiter. (Note that our table does not show all of the cumulative probabilities through the week, although we could easily add a column for that.)

Do you remember the second question? Here it is again: Today is Monday: If you go to the restaurant today who is most likely to be your server? What is the specific probability that your server will be that person? At this point you should be able to follow the same procedure shown above, only this time you will isolate one row instead of one column. Specifically, you should isolate the Monday row and use the marginal total for Monday (0.1290) to normalize the probabilities in that row. You should do it yourself, but here is a crosscheck for your answer: if it is Monday there is a probability of 0.421 that your server is Devon.

CONCLUSION

You may remember at the beginning of this chapter that I posed this question: "When we are presented with a new piece of information about a probability scenario, how does it change what we believe?" You should now be able to see that adding a piece of information, such as what server we have or what day it is, allows us to refine our understanding of what is going on. We can use our **prior** understanding of the probabilities of certain outcomes in combination with a new piece of information or evidence to develop a more highly refined **posterior** set of probabilities. This kind of thinking is at the heart of the Bayesian statistical techniques that are now coming into common practice. Over the next few chapters, when we look at the logic of statistical inference, we will use this Bayesian thinking as one method of considering what our evidence tells us about the relative likelihoods of different outcomes.

Another important message of this chapter, though, is that reasoning about probability can be easier to comprehend if we think about a set of real events rather than just a bunch of decimals or fractions. We can use a table of event outcomes, marginal totals, and normalization to create displays that help us with our reasoning. Anytime you are given a probability problem to think about, try turning it into a set of 100 or 1,000 events and then breaking things down according to the various possible outcomes. Many people find it easier to think about a specific number of events out of 100 or 1,000 rather than thinking about decimal probabilities like 0.2632 or 0.421. How about you?

EXERCISES

1. Flip a fair coin nine times and write down the number of heads obtained. Now repeat this process 100,000 times. Obviously you don't want to have to do that by hand, so create the necessary lines of R code to do it for you. Hint: You will need both the rbinom() function and the table() function. Write down the results and explain in your own words what they mean.

2. Using the output from Exercise 1, summarize the results of your 100,000 trials of nine flips each in a bar plot using the appropriate commands in R. Convert the results to probabilities and represent that in a bar plot as well. Write a brief interpretive analysis that describes what each of these bar plots signifies and how the two bar plots are related. Make sure to comment on the shape of each bar plot and why you believe that the bar plot has taken that shape. Also make sure to say something about the center of the bar plot and why it is where it is.

3. Using the data and results from Exercises 1 and 2, create a bar plot of *cumulative* probabilities. Write an interpretation of that bar plot as well, making sure to explain how cumulative probabilities differ from the results you described in Exercise 2.

4. Run 50 trials in which you flip a fair coin four times. As a hint, you should use rbinom() and table() together. Save the results or write them down. Then, using the "prob=0.8" argument with the rbinom() command, run another set of 50 trials (with four flips) with an "unfair" coin (i.e., a coin where the probability of heads is not 0.5). Next, using a piece of paper (or appropriate software) write down the results of the fair coin as the first row of a contingency table and the results of the unfair coin as the second row of the table (each row should have five entries, not counting the marginal row total). Calculate

marginal totals and the grand total and write these in the appropriate places on your contingency table. Forget the R commands for a minute, and explain what aspects of your contingency table (in other words, which elements of data in the table) show that the unfair coin is unfair.

5. Convert all of the cells in the contingency table that you developed for Exercise 4 (including the marginal totals and the grand total) into probabilities. Explain in your own words which cell has the highest probability and why.

6. One hundred students took a statistics test. Fifty of them are high school students and 50 are college students. Eighty students passed and 20 students failed. You now have enough information to create a two-by-two contingency table with all of the marginal totals specified (although the four main cells of the table are still blank). Draw that table and write in the marginal totals. I'm now going to give you one additional piece of information that will fill in one of the four blank cells: only three college students failed the test. With that additional information in place, you should now be able to fill in the remaining cells of the two-by-two table. Comment on why that one additional piece of information was all you needed in order to figure out all four of the table's main cells. Finally, create a second copy of the complete table, replacing the counts of students with probabilities. What is the pass rate for high school students? In other words, if one focuses only on high school students, what is the probability that a student will pass the test?

7. In a typical year, 71 out of 100,000 homes in the United Kingdom is repossessed by the bank because of mortgage default (the owners did not pay their mortgage for many months). Barclays Bank has developed a screening test that they want to use to predict whether a mortgagee will default. The bank spends a year collecting test data: 93,935 households pass the test and 6,065 households fail the test. Interestingly, 5,996 of those who failed the test were actually households that were doing fine on their mortgage (i.e., they were not defaulting and did not get repossessed). Construct a complete contingency table from this information. Hint: The 5,996 is the only number that goes in a cell; the other numbers are marginal totals. What percentage of customers both pass the test and do not have their homes repossessed?

8. Imagine that Barclays deploys the screening test from Exercise 6 on a new customer and the new customer fails the test. What is the probability that this customer will actually default on his or her mortgage? Show your work and especially show the tables that you set up to help your reasoning.

9. The incidence of HIV in the U.S. population is 0.00036 (0.036%). There is a blood test for HIV, but it is not perfect in its ability to detect the disease. If a person has HIV, the probability is 89% that the blood test will be "positive" (meaning that it has detected the disease). If a person does not have HIV, the probability is still about 7% that the blood test will nonetheless be positive. Imagine a person who gets the blood test and learns that the test came out positive. What is the probability that this person actually has HIV? Remember what the chapter said about turning probability problems into lists of real events (I would suggest using 100,000 events in this case because of how small the incidence of HIV is). Show your work and especially the contingency tables that you set up to help your reasoning.

CHAPTER 3

Probabilities in the Long Run

In the previous chapter we focused on specific scenarios where a set of events occurred, we summarized them in a table or a graph, we converted them into probabilities, and we interpreted the results. The thinking behind this work was that we would simply take events as we found them and not worry too much about whether those events were *representative* of a larger realm of possibilities. Since the beginning of the development of statistics in the 1700s, statisticians have also thought about ideal models that may exist behind the scenes of a given phenomenon. Carrying forward our example from the last chapter: What if we could develop a statistical model that adequately represented a process such as toast dropping? We could use that model to generate a summary of long-range outcomes and then compare whatever real data we collect to those ideal outcomes. That could tell us something useful about the data we collected: Do our new data fit the ideal model or not? If they do fit the model, then we may have confirmed our assumptions; if they don't fit the model, something important may have changed.

So let's pretend that deep within the physics department of our local university, using complicated machinery and powerful computers, scientists have created an ideal model of the angle of impact of pieces of toast. If you remember a little of your trigonometry, we can say that some toast lands perfectly flat with an angle of 0 degrees while other toast also lands flat (but on the other side with an angle of 180 degrees). Naturally, lots of toast strikes the floor at all of the angles between 0 and 180 degrees. If we ignore the possibility that toast might bounce or stand on edge, we can assert that any toast landing at an angle between 0 and 90 will settle with one side down and every other piece of toast will land with the other side down. If we had a data set showing the impact angles of lots of pieces of toast, we might be able to get a better understanding of when and why the jelly side lands down or up.

Our friendly physicists have provided an R command to generate just such a set of toast drops from this ideal process. Here's the code for generating the toast-drops data and looking at a histogram of that data set:

```
# Random numbers from uniform distribution
toastAngleData <- runif(1000,0,180)
head(toastAngleData)      # Look at the first few numbers in the list
tail(toastAngleData)      # Look at the last few numbers in the list
hist(toastAngleData)      # Plot a histogram of all 1000 data points
```

By the way, runif() creates random numbers in the so-called **uniform distribution.** When you look at the histogram, what do you think "uniform" signifies in this context? In addition, based on your review of Chapter 1 and your inspection of the histogram, you should be able to calculate a mean of this data set on your own. Before you run that command, however, what do you predict that the mean will be? That mean will be important later on in this process. For now, though, I want to introduce an important new process that is essential to statistical thinking. The process is called **sampling.**

SAMPLING

Toast and gumballs do not generally go together, but just for a moment imagine a jar full of gumballs of two different colors, red and blue. The jar was filled from a source that provided 100 red gum balls and 100 blue gum balls, but when these were poured into the jar they got all mixed up. If you drew eight gumballs from the jar at random, what colors would you get? If things worked out perfectly, which they hardly ever do, you would get four red and four blue. This is half and half, the same ratio of red and blue that is in the jar as a whole. Of course, it rarely works out this way, does it? Instead of getting four red and four blue you might get three red and five blue or any other mix you can think of. In fact, it would be possible, though perhaps not likely, to get eight red gumballs.

The basic situation, though, is that we really don't know what mix of red and blue we will get with one draw of eight gumballs. The process of drawing a subset of elements from a "master" set—often referred to as a **population**—is known as **sampling.** The fact that we don't get exactly half red and half blue in many cases is known as **sampling error.** Sampling error occurs to varying and unknown degrees when a statistic obtained from a sample (our statistic in this example is the proportion of red gumballs) does not precisely match the **parameter** from the population. In our gumball jar example, the population comprises the 200 gumballs and the population parameter is 50% red (or 50% blue, if you prefer). When we draw gumballs out of the jar, we can estimate the population parameter from the proportions of red and blue gumballs in our sample. That estimate, the statistic, typically will not match the population parameter. In our gumball example, we happen to know that the population parameter is 50% red, because we filled the jar (the population) ourselves, but in real-world research we usually do not know the population parameter and therefore cannot directly ascertain the amount of sampling error that arises from any particular sample.

Many different kinds of sampling exist, but in this book we will mainly focus on **random sampling with replacement.** For mathematical reasons, this is one of the easiest kinds of sampling to consider. There's really just one rule that summarizes the process of random sampling with replacement and that is that the next draw of an element has an equal chance of pulling out any member of the population. Thinking about our jar of gumballs, that means that on any given draw every single gumball has exactly $1/200 = 0.005$ probability of being selected.

Many learners find the "with replacement" part of the process a bit strange. Think about it this way: If you are only drawing out one gumball and you have an excellent process for mixing them up and choosing one at random, then you indeed have an equal chance of picking any given gumball. What gets complicated is the second draw: If you put the first gumball in your mouth while you are pulling out the second one, then there is a zero chance that you will draw that first gumball, whereas now the remaining ones in the jar each have a $1/199 = 0.005025$ chance of being selected. Although that is not a big difference from the first draw, if we keep going, the probabilities get very messy. For that reason, sampling with replacement means that after each draw of a gumball, you throw back the gumball you selected (and, if necessary, mix them all up again) to ensure that the next draw has the same probabilities as the first draw. Make sure to wear rubber gloves so that you don't get your germs all over the gumballs you throw back in the jar.

Let's consider another example just to get the hang of it. I once lived in a place called Watertown, which currently has a population of 31,915. By the way, statisticians use a capital N to refer to the number of elements in a population, so for Watertown, $N = 31,915$. The nice folks at the Watertown library want to conduct an opinion survey of Watertown residents, to ask them about how much they value the library on a scale of 0 to 10. The librarians decide that if they got 50 people to respond to the survey, they would learn what they needed to know. In other words, when they calculate the mean of their sample of 50 people (on the scale of 0 to 10), they hope that statistic will adequately reflect the population parameter—what you would get if you asked all 31,915 Watertown residents the question.

Statisticians use a small letter n to refer to the number of elements in a sample, so the librarians will obtain a sample of $n = 50$. To obtain a truly random sample, the librarians would need to begin somehow with a complete contact list of all of Watertown's residents, and then they would need to choose $n = 50$ names from the list at random. If they sample with replacement, as described above, there is a possibility that a person will get chosen to take the survey more than once, but that possibility is extremely small, so the librarians don't worry about it.

One diligent library intern receives the list of randomly chosen residents and drives around to residents' houses asking the opinion question. Some people may have moved and would therefore be considered **unreachable.** Other people come to the door, but don't want to answer the questions: they are **refusals.** Generically speaking, anyone who is sampled but who does not provide a response is a **nonrespondent.** To the extent that there are nonrespondents from the original randomly selected list of $n = 50$, there is the possibility of nonresponse **bias.** Nonresponse bias would occur if the sampled people who failed to participate were different in some important way from those who did participate. Let's say, however, that the intern was immensely persistent (and lucky!) and obtained surveys from all 50 of the sampled individuals. We know from our earlier gumball example that the $n = 50$ in the sample probably do not perfectly represent the $N = 31,915$ residents of Watertown.

Now Watertown library is lucky enough to have two interns working there this year. So the librarians decide to sample another $n = 50$ residents and send the second intern out to collect an additional sample of survey responses. Now we have two different samples of opinion data obtained from the residents of Watertown. The two sample means (on the 0 to 10 scale) from these two sets of data will not match each other perfectly, although they may be close to each other. Here's a crazy notion, though: What if Watertown had lots of interns—12 or 24 or 48 or 96 interns—and each one went out and obtained their own random sample of data? If we lined up all of the means from these dozens of different interns' samples, we would be able to get a flavor for how much variation there was among random samples of size $n = 50$. This is precisely what statisticians have figured out how to do with their mathematical models and what we will simulate below with the help of R.

To review: the process of repeatedly drawing elements from a population is called "sampling." Once we have decided on a sample size we can repeatedly draw new samples of that size from a population. This repetitious process is practically impossible to do in real life, even with an army of interns, but we can do a simulated version of it on the computer with the help of R. For every sample we draw, we will measure some aspect of it (such as the mean of a survey question) and record that measurement to create what is known as a **sampling distribution.** You know the concept of a distribution from Chapter 1: we are using the same concept here to describe a collection of data that comes from the process of repeatedly sampling.

REPETITIOUS SAMPLING WITH R

Keeping in mind that this is a process nearly impossible to accomplish in actual research, let's return to our toastAngleData to illustrate the process of sampling repeatedly from a population. The goal of doing this, as explained earlier, is to get a feel for how much variation happens because of sampling error. As always, R provides a convenient function to draw a sample from a population, sensibly called sample(). Let's begin by drawing a small sample from toastAngleData.

```
sample(toastAngleData,size=14,replace=TRUE)
```

This command asks R to draw a sample of size $n = 14$ from the toastAngleData (which we are treating as a population) and to use sampling with replacement as described above. Here's the output I got:

```
[1] 152.07620 102.20549 89.35385 42.75709 21.13263 158.35032 141.95377
[8] 16.31910 136.32748 171.54875 44.39738 36.60672 87.92560 149.97864
```

Notice that we have some small angles, such as 16.3, and some large angles such as 171.5. You can run that same sample() command a few different times to show that you get a different random sample of data each time. Occasionally, you may notice the exact same angle appearing in your sample twice. Remember that using sampling with replacement, this is an expected outcome. We can summarize each sample that we draw from the data set by running it into the mean() function, like this:

```
mean(sample(toastAngleData,size=14,replace=TRUE))
```

When I ran this command I got a mean value of about 93.8. Think about what this signifies. Remember that the toast can fall at any angle between 0 and 180 degrees, and that we would assume, given no evidence to the contrary, that the angles of impact will cover that whole range uniformly. So, it should make perfect sense to you that the mean of any given sample is quite likely to be close to 90 degrees. That's the halfway point between 0 and 180 degrees. Also, remember before when I said that we would measure some aspect of each sample that we drew? The mean of the 14 toast angles is the aspect of the sample that we are measuring.

Now keep in mind that unlike the librarians in the previous example, we are not mainly interested in any one sample or its particular mean. We know that most individual samples have sampling error such that they do not exactly match the population. Instead we want to use this simulation to see what happens across a lot of samples if we look at them over the long haul—this will give us a nice picture of what amounts of sampling error we can expect.

We want to get R to repeat the sampling process for us, not once, not four times, but hundreds or thousands of times. Like most programming languages, R has a variety of ways of repeating an activity. One of the easiest ones to use is the replicate() function. To start, let's just try four replications:

replicate(4, mean(sample(toastAngleData,size=14,replace=TRUE)))

Here's the output that I obtained from this command:

[1] 82.15252 98.11379 77.42944 97.23404

Your results will be different, and every time you run this command you will get a slightly different result because the sample() command is choosing from the toast angle data at random. Note that unlike my first run of the process, these four means are not too close to 90, although they are still in the neighborhood. While things are still nice and simple here, it would be a good time to pause, reflect, and make sure you understand what is happening, both conceptually and with the specifics of this R command. In the command just above, we have asked R to repeat a command four times and show us a list of the results. Each time that R runs one of these replications, it executes this command:

mean(sample(toastAngleData,size=14,replace=TRUE)).

That command draws a sample of size $n = 14$ from our population of toast impact angles and calculates the mean of the 14 data points (angles) in the sample. The result is a vector of means—one mean to represent each sample—and naturally at this point the list comprises just four means.

If that all makes sense to you, let's crank it up to the next level. Instead of running just four replications, let's run 10,000 and make a histogram of what we get:

```
samplingDistribution <- replicate(10000,mean(sample(toastAngleData,
    size=14,replace=TRUE)),simplify=TRUE)
hist(samplingDistribution)
```

The code shown above runs 10,000 replications—which creates a vector of 10,000 sample means—and places it in a storage location called "samplingDistribution." Then we create a histogram of samplingDistribution. If you run this code, you will get something that looks very much like Figure 3.1.

I'm sure that you quickly noticed the shape of this distribution. With that characteristic bell shape, it closely matches the normal curve that we first encountered in Chapter 1. Furthermore, the curve is centered on 90 degrees. You can also reason pretty quickly from this that it is exceedingly unlikely to draw a sample of 14 toast impact angles that has a mean as low as 40 or as high as 140. Although these strange and extreme samples do occur from time to time, they are quite rare. Next, we are going to take the interesting step of calculating a **grand mean** from our long list of 10,000 individual sample means. Let's find out the grand mean of this sampling distribution and compare that to the mean of our original (raw) population data:

```
# The mean of our 10,000 different sample means
mean(samplingDistribution)
mean(toastAngleData)      # The mean of the raw angle data
```

The two values I obtained from this code were 90.5 for the mean of the sampling distribution and 90.6 for the mean of the original raw data. Not exactly identical, but close to within about 0.1%. Keep your eye on those two commands: it is really extremely important that you understand the different ways that those two means were calculated. You may want to draw a diagram on a sheet of paper to prove to yourself that you understand where the two means came from: draw a big blob for the population of toast angles, some squares to represent a few of the individual samples, and arrows to connect everything together. To reiterate the explanation, the first command, mean(samplingDistribution), calculates the grand mean of a list of 10,000 means, where each of those 10,000 means summarizes one sampling event from a random sample draw of $n = 14$ toast angles. The second command, mean(toastAngleData), goes all the way back to the original 1,000 raw data points that we created right at the beginning of the chapter. That second command calculates the raw data mean of the

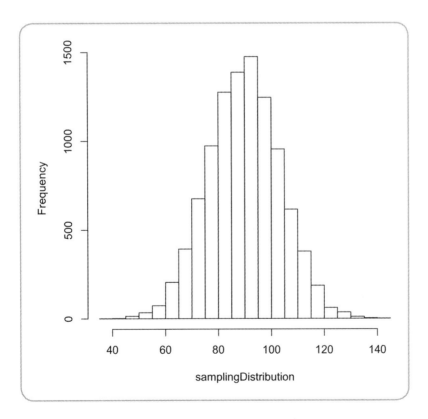

FIGURE 3.1. Histogram of sampling distribution for $n = 10,000$ replications of calculating the mean from a randomly drawn sample of $n = 14$ angles.

whole population of toast angles: the population mean. Go over this distinction in your mind a few times: the grand mean calculated from a set of means versus the population mean.

The fact that these two means are so similar to one another is very "meaningful" for us (sorry!). Based on the fact that these two means are so close, we can conclude that although any one sample of 14 angles will probably not have a mean that is exactly 90 degrees (and a very few samples will have means that are quite low or high), over the long run, *the mean of a sampling distribution matches the mean of the original population data.* We can also say that most of the sample means cluster quite nearby to the population mean but that there are a few samples that are out in the "tails"—in other words, very far from the actual population parameter.

Now is a great moment to take a deep breath. We have just covered a couple hundred years of statistical thinking in just a few commands. In fact, there are two big ideas here, the **law of large numbers** and the **central limit theorem** that we have just partially demonstrated. These two ideas literally took mathematicians like Gerolamo Cardano (1501–1576) and Jacob Bernoulli (1654–1705) several centuries to figure out (Fienberg, 1992). Of course, without R they had to count on their fingers and that took time. Also, ink and parchment were really expensive back then. If you look these ideas up, you will find a good deal of mathematical details, but for our purposes, there are two incredibly important take-away messages. First, if you run a statistical process like sampling a large number of times, it will generally converge on a particular result. This is what Bernoulli postulated more than 300 years ago: run an experiment such as a coin toss a large number of times, and the proportion of heads will converge on the true value (50% for a fair coin) as the number of trials gets larger (Gorroochurn, 2012).

Second, when we are looking at statistics such as sample means, and when we take the law of large numbers into account, we find that a distribution of sampling means starts to create a bell-shaped or normal distribution. The center of that distribution, that is, the mean of all of those sample means, ends up really close to the actual population mean. This is a powerful idea, originally attributed to 19th-century scientist Pierre-Simon LaPlace, which offers us important knowledge about the nature of sampling error (Fischer, 2010, p. 18). Remember that the normal curve tends to appear whenever something we measure contains small variations that mostly cancel each other out most of the time. That's exactly what a sample is: we draw elements from a population, some are high and some are low, and when we calculate a sample mean, we mostly cancel out the highs and lows to get a result that is close to but not identical to the population mean. When we repeat that sampling process many times we get quite a lot of sample means near the population mean and relatively few that are far from the population mean. A histogram of those sample means will look more and more like a bell curve as we repeat the process more and more times. Next, we will take advantage of the shape of the resulting bell curve to reason more precisely about how much sampling error we are likely to observe.

USING SAMPLING DISTRIBUTIONS AND QUANTILES TO THINK ABOUT PROBABILITIES

Earlier, I suggested that it was exceedingly unlikely to draw a sample of 14 toast-impact angles that had a mean angle as low as 40 or as high as 140. Just how unlikely is it? Furthermore, if we want to reason about other mean values that we might encounter, such as 100 or 175, what are the likelihoods of finding values like those? It turns out to be pretty straightforward to reason out this kind of thing just by looking at the histogram, but we also want to get some help from precise calculations that R can do for us. Let's go back to the simple summary() command that R provides for summarizing any vector of numeric values.

summary(samplingDistribution)

Here's the output that my command generated. Yours will be slightly different, but should be very close to these numbers:

Min.	1st Qu.	Median	Mean	3rd Qu.	Max.
35.98	81.32	90.41	90.46	99.84	140.20

By now you should have no problem interpreting min and max, as well as mean and median. For the moment, we are more interested in the quartiles. **Quartile** indicates that we are dividing up a set of values into four equal parts: one quarter on the low end, one quarter on the high end, and two quarters in the middle. You should be able to make sense of this definition by going back to Figure 3.1. For example, from the output of summary() the first quartile is at 81.32. That value signifies that 25% of all sample means we drew from the population have values *at or below 81.32*. Likewise, looking at the third quartile, 25% of sample means have values *at or larger than 99.84*.

Let's test your understanding: What is the proportion of sample means that falls between 81.32 and 99.84? You should easily be able to answer that based on the definition of quartiles and the output displayed above. If you need another hint, 25% of cases are between 81.32 and the median, and another 25% of cases are between the median and 99.84. Put it all together and it should be evident that half of all the cases in a set of values fall between the first quartile and the third quartile.

Just as an aside, here's a neat concept from calculus: we can examine the "area under the curve" in order to estimate or ascertain the proportion of cases above, below, or between certain values on the X-axis. The use of the word *curve* might bother you a bit, because the histogram in Figure 3.1 shows discrete categories of data rather than a continuous curve that you might get from plotting a function. Even so, we can get a good deal done by just pretending that we are looking at a smooth, continuous curve and focusing on the area between the curve and the X-axis. You should look closely at Figure 3.1, see where the first quartile, median, and third quartile are, and visually confirm for yourself that these three cut-points divide the area under the curve into four equal pieces.

While quartiles are fun and easy because most of us are used to dividing things into fourths, we also need to be able to make finer cuts through our curve. R provides us with a function called quantile() to divide things up into an arbitrary number of divisions:

```
quantile(samplingDistribution, c(0.01, 0.05, 0.50, 0.95, 0.99))
```

It can be hard to remember the difference between quartile and quantile, but just remember that "quart" is a quarter of something and "quant" (meaning "number") is cutting something into any number of pieces. The first argument to quantile() is our vector of sampling means. The second argument provides a list of the quantiles or cut-points that we would like R to use over these data. Look carefully at the items in this list. The one in the middle, .50, should be pretty sensible to you: that is the median and it should be identical or very close to what the summary() function produced. The other values in the list might seem less familiar. I have specifically and intentionally asked for cut-points in the extreme right- and left-hand **tails** of the distribution of sample means. The tail of a curve is defined as either the lowest or highest valued group of items in a distribution—reaching all the way out to include all of the extreme values. For the normal distribution, which goes down in a nice smooth pattern on the left and the right, you might think of the tail as looking like the tail of a dinosaur. You specify where the tail starts by choosing a cut-point—for example, the right-hand tail starting at the 95th percentile.

Here is the output that the quantile() function produced:

1%	5%	50%	95%	99%
58.50513	67.92841	90.41403	112.88387	121.98952

So, just as we predicted, the 50th percentile is the median and it falls right around 90. Just 1% of all of the sample means are at or below a value of 58.51. Likewise, just 1% of all of the sample means are at or above a value of 121.99. Similarly, just 5% of all of the sample means are at or below a value of 67.93, while 5% of all of the sample means are at or above a value of 112.88. We can easily show this result on an updated graph using the abline() function to put some cut-off lines on the histogram:

```
hist(samplingDistribution)
abline(v=quantile(samplingDistribution,0.01))
abline(v=quantile(samplingDistribution,0.05))
abline(v=quantile(samplingDistribution,0.95))
abline(v=quantile(samplingDistribution,0.99))
```

The abline() function draws a line on an existing graph from point a to point b, but it also contains some shortcuts that make it easy to do horizontal and vertical lines. The v parameter to abline() requests a vertical line at the point on the X-axis that is specified. Note that my four commands to abline() repeat the calls to quantile() so that when you run them you will get the correct values for your sampling distribution. Also note that I did not bother to put in a line for the median since you already know where that should appear. The resulting histogram appears in Figure 3.2.

Spend some time staring at this graph so that you can make sense of the vertical lines—our different cut-points for the tails. See how nice and symmetrical the histogram looks. The 1% and 99% lines in particular highlight how few of our sample means fell into these extreme zones. In fact, we could easily do a little more exploration of the data that would highlight just what we are working with. Let's look more closely at the bottom 1% and then leave the others for you to do as an exercise. Remember that there are 10,000 sample means in our data set, so 1% should be about 100 sample means. We can ask R to just let us look at that set as follows:

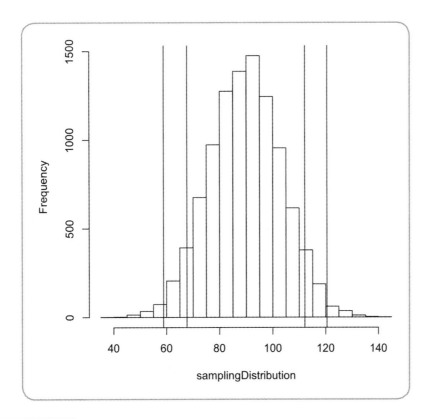

FIGURE 3.2. Histogram of sampling distribution with 1%, 5%, 95%, and 99% tails indicated with vertical lines.

```
samplingDistribution[ samplingDistribution <= quantile(samplingDistribution,
    0.01) ]
```

In R the left square brace [and the right square brace] are what R programmers call "subsetting operators," and they allow us to specify particular elements of a vector. Inside the square braces I have placed a logical condition, namely, that I only want to see the elements of the vector with values at or below the 0.01 quantile. This is our first encounter with both subsetting and logical conditions, so it makes sense to pause and make sure you understand what this line of code accomplishes. Keeping in mind that samplingDistribution is a vector, R evaluates the logical condition for each element of the vector and only reports those elements where the condition is true. The logical condition is true in each case where an element of the vector is less than or equal to the 0.01 quantile of the distribution.

If you run that one powerful line of code you will indeed get a vector of 100 sample means just as we predicted. I will not waste space by reproducing that whole list here, but I will run a summary() on the result so that we can have an overview:

```
summary( samplingDistribution[samplingDistribution <= quantile
    (samplingDistribution,.01)] )
```

The output of this command shows that the maximum value in my subset was 58.4: right underneath the 1% quantile that R reported earlier and that is plotted as the left-most vertical line in Figure 3.2. That shows that we have indeed peeled off the bottom 100, or exactly 1% of the sample means. You might also note that the minimum value of 35.98 is the smallest mean in the whole data set, which you can verify by going back to the beginning of this chapter and looking at the results of the summary() function when we ran it on the whole list of sample means.

Now all of this seems like a lot of trouble, so why bother, you might be asking yourself? We want to know about extreme examples because they are quite unlikely and so it is very atypical for us to observe them. We can use this result to reason about *new* examples that we observe. For example, if we ran a new sample of $n = 14$ toast-drop angles and we found that the sample mean was an angle of about 25, what would you say about that sample mean? I'm hoping you would say that it is wayyyyyyyy extremely low, because you would be correct. The reason you know that you are right and I know that you are right is that we have the big vector of sampling means to compare with and that vector of sampling means makes a pretty good bell curve when we look at a histogram of it. We can use Figure 3.2 to state quite strongly that any sample with a mean that is less than about 58.5 is pretty darned low—lower than 99% of the other means in our vector of 10,000 sample means.

CONCLUSION

In this chapter, we went totally hickory nuts with the toast example by creating a fake population of angles of toast drop. Presumably, if we were really doing research in this area, the distribution of angles of toast drop would let us understand the jelly-up/jelly-down situation better. Once we obtained our fake population, we started to draw random samples from it, lots and lots of them, to create a sampling distribution of means. Each random sample had a mean value that was like a mini-snapshot of the data set. Very few, if any, of these snapshots showed the exact same mean as the mean of the initial data set. That's what happens when you randomly sample. Rarely do you get a sample that perfectly matches the population. At the same time, though, it is also rare to get a sample with a mean that is way higher than or way lower than the population mean.

We plotted a histogram of a large set of sample means and found that the distribution of sampling means formed a normal curve. By using quantiles, we could designate different regions of this normal curve and understand what proportion of the sample means fell into each of these regions. Finally, given our detailed understanding of these regions of the sampling distribution we could look at a new piece of information, that is, a new sample mean that somebody handed us, and tell something about how likely or unlikely it would be to observe such a sample mean.

To get all of that work done, we needed the runif() command to generate a list of random numbers in a uniform distribution. We also used the sample() command to draw a random sample of values from our list. In another new development, we used a function called replicate() to do a job over and over again, in this case repeatedly running the sample() command. Just as a point of trivia, replicate() is what R programmers call a **wrapper function.** A wrapper function hides the work of a complex activity by putting a simpler user interface around it. Before moving onto the next chapter, you should make sure to experiment with runif(), sample(), and replicate() so that you are comfortable doing new activities with them. You should also practice your use of the subsetting operator (the square brackets) and logical conditions.

In the next chapter, we use the same kind of thinking as we did here to formally introduce the logic of statistical inference. Before you move on to the next chapter, you should make sure that you can use R to produce your own distribution of sampling means, graph it, and figure out where the cut-off points are for extremely low and extremely high values. With a pencil and paper or on a white board you can also practice drawing normal curves and showing on those curves where the median is, and where the extreme low tail and the extreme high tail are on the curve.

EXERCISES

1. Marlboro College in Vermont is one of the smallest liberal arts colleges in the United States, with an enrollment of about 240 students. Consider this whole group to be the population of interest for a study of the heights of students. In a few sentences, describe how you would draw a random sample of students from this population and measure each person's height. Choose a sample size that seems practical, given your proposed methods. Make sure to mention what is the population parameter of interest and what is the sample statistic. When your data collection is done, will the sample statistic perfectly match the population parameter? Explain why or why not.

2. For the remaining exercises in this set, we will use one of R's built-in data sets, called the "ChickWeight" data set. According to the documentation for R, the ChickWeight data set contains information on the weight of chicks in grams up to 21 days after hatching. Use the summary(ChickWeight) command to reveal basic information about the ChickWeight data set. You will find that ChickWeight contains four different variables. Name the four variables. Use the dim(ChickWeight) command to show the dimensions of the ChickWeight data set. The second number in the output, 4, is the number of columns in the data set, in other words the number of variables. What is the first number? Report it and describe briefly what you think it signifies.

3. When a data set contains more than one variable, R offers another subsetting operator, $, to access each variable individually. For the exercises below, we are interested only in the contents of one of the variables in the data set, called weight. We can access the weight variable by itself, using the $, with this expression: ChickWeight$weight. Run the following commands, say what the command does, report the output, and briefly explain each piece of output:

```
summary(ChickWeight$weight)
head(ChickWeight$weight)
mean(ChickWeight$weight)
myChkWts <- ChickWeight$weight
quantile(myChkWts,0.50)
```

4. In the second to last command of the previous exercise, you created a copy of the weight data from the ChickWeight data set and put it in a new vector called myChkWts. You can continue to use this myChkWts variable for the rest of the exercises below. Create a histogram for that variable. Then write code that will display the 2.5% and 97.5% quantiles of the distribution for that variable. Write an interpretation of the variable, including descriptions of the mean, median, shape of the distribution, and the 2.5% and 97.5% quantiles. Make sure to clearly describe what the 2.5% and 97.5% quantiles signify.

5. Write R code that will construct a sampling distribution of means from the weight data (as noted above, if you did exercise 3 you can use myChkWts instead of ChickWeight$weight to save yourself some typing). Make sure that the sampling distribution contains at least 1,000 means. Store the sampling distribution in a new variable that you can keep using. Use a sample size of $n = 11$ (sampling with replacement). Show a histogram of this

distribution of sample means. Then, write and run R commands that will display the 2.5% and 97.5% quantiles of the sampling distribution on the histogram with a vertical line.

6. If you did Exercise 4, you calculated some quantiles for a distribution of raw data. If you did Exercise 5, you calculated some quantiles for a sampling distribution of means. Briefly describe, from a conceptual perspective and in your own words, what the difference is between a distribution of raw data and a distribution of sampling means. Finally, comment on why the 2.5% and 97.5% quantiles are so different between the raw data distribution and the sampling distribution of means.

7. Redo Exercise 5, but this time use a sample size of $n = 100$ (instead of the original sample size of $n = 11$ used in Exercise 5). Explain why the 2.5% and 97.5% quantiles are different from the results you got for Exercise 5. As a hint, think about what makes a sample "better."

8. Redo Exercises 4 and 5, but this time use the precip data set that is built into R. The precip data only includes one variable, so you will not need to use the $ subsetting operator. Use a sample size of $n = 1,000$. Hint: When you use the sample() command, make sure to include "replace=TRUE" as part of the command. Comment on why, for this particular data set, it is a curious idea to choose a sample size of $n = 1,000$. Also, explain why this does actually work.

## CHAPTER 4

Introducing the Logic of Inference Using Confidence Intervals

"Inference" refers to a reasoning process that begins with some information and leads to some conclusion. If you have ever taken a logic course or thought about logical reasoning, you have probably heard things like, "All mammals are warm-blooded animals; this animal is a mammal; therefore this animal is warm-blooded." That particular sentence is known as a "syllogism" and it represents a kind of inference called "deduction." Another kind of inference, **induction**, reasons from specific cases to the more general. If I observe a cat jumping from a tree and landing on its feet, and then I observe another cat, and another, and another doing the same thing, I might infer that cats generally land on their feet when jumping out of trees. Statistical inference takes this same kind of logical thinking a step further by dealing systematically with situations where we have uncertain or incomplete information. Unlike the syllogism about warm-blooded animals presented above, conclusions that we draw inductively from samples of data are never fixed or firm. We may be able to characterize our uncertainty in various ways, but we can never be 100% sure of anything when we are using statistical inference. This leads to an important idea that you should always keep in mind when reasoning from samples of data:

**You cannot *prove* anything from samples
or by using statistical inference.**

I emphasize this point repeatedly with my students and I expect them to know the reason why (by the end of the course if not before). So, if you ever hear a journalist or a scientist or anyone else saying that statistical analysis of one or more samples of data *proves* a certain conclusion, you can be assured that he or she is mistaken, and perhaps misinformed or being intentionally misleading.

Instead of setting out to prove something, we collect data and analyze it with inferential statistics in order to build up a weight of evidence that influences our certainty about one conclusion or another. There's a great historical example of this method from the 19th-century medical researcher John Snow. Snow (1855) studied outbreaks of cholera in London. Cholera is a devastating bacterial infection of the small intestine that still claims many lives around the world. John Snow mapped cases of cholera in London and found that many cases clustered geographically near certain wells in the city where residents drew water for drinking and other needs. Using this purely graphical method, Snow was able to infer a connection between the cholera cases, the wells, and sewage contamination that led him to conclude that fecal contamination of drinking water was the primary mechanism that caused cholera. Blech! His map, his evidence, and his reasoning were not a proof—in fact, many authorities disbelieved his proposed mechanism for decades afterward—but Snow's work added substantially to the weight of evidence that eventually led to a scientific consensus about cholera infections.

Today we have many more data analysis tools than John Snow did, as well as an unrivaled ability to collect large amounts of data, and as a result we are able to more carefully quantify our ideas of certainty and confidence surrounding our data sets. In fact, while there was one predominant strategy of statistical inference used during the 20th century, known as "frequentist statistical inference," some new approaches to inference have come into play over recent years. We will begin by considering one element of the frequentist perspective in this chapter and then add to our knowledge by considering the so-called Bayesian perspective in the next chapter.

But now would be a really good point to pause and to look back over the previous chapter to make sure you have a clear understanding of the ideas around sampling distributions that I introduced there. In the previous chapter we began with a randomly generated data set that we used to represent the whole population—specifically a whole population of angles at which toast struck the ground after being dropped off a plate. It is important to now declare that this made-up example was ridiculous on many levels. First of all, I hope that there are not many scientists who spend their time worrying about toast-drop angles.

Most importantly, however, as researchers *we almost never have access to the totality of the population data*. This is a critical conceptual point. Although in some special cases it may be possible to access a whole population (think, e.g., of the population of Supreme Court justices), in most research situations it is impractical for us to reach all of the members of a population. Think about the population of humans on earth, or of asteroids in the asteroid belt, or of tablet computers deployed in the United States, or of electric utility customers in Canada. In each case we can define what we are talking about, and we can have criteria that let us know whether a particular person, or tablet, or asteroid is part of the population, *but we cannot get data on the complete population*. It is logistically impossible and/or cost-prohibitive to do so. Even in smaller populations—the students enrolled in a college, all of the employees of a company, the complete product inventory of one store, a full list of the text messages you have ever sent—there may be good, practical reasons why we may not want to measure every single member of a population.

That's why we sample. Drawing a sample is a deliberate act of selecting elements or instances such that the collection we obtain can serve as a stand-in for the whole population. At its best, a sample is *representative* of a population, where membership in a population can be defined but the elements of the population can never be accessed or measured in their totality. As we know from our activities in the previous chapter, it is rare to ever draw a sample that perfectly mimics the population (i.e., one particular sample that has the exact same mean as the population). In this context, the goal of statistical inference is to use a sample of data to make estimates and/or inferences about a population, with some degree of certainty or confidence. (Note that in those unusual cases where we can measure every element of a population, we conduct a **census** of the population and we do not need sampling or inferential thinking at all. Whatever we measure about all Supreme Court justices is what it is, within the limits of measurement precision.)

Let's now turn to a real-world example of populations and samples to illustrate the possibilities. We will use another built-in data set that R provides called "mtcars." If you type "?mtcars" at the command line, you will learn that this is an old (1974!) data set that contains a sample of 32 different vehicles with measurements of various characteristics including fuel economy. I'll bet you are happy that I have finally stopped talking about toast! Historical point of trivia: cars from the 1974 model year were the last ones that were designed prior to the 1970s oil crisis (in most parts of the world gas prices tripled between 1974 and 1980), so the average fuel economy of these cars is shockingly low. In this chapter, we will focus on one simple research question: Do cars with automatic transmissions get better or worse mileage than cars with manual transmissions? As usual, I provide R code below to show what I am doing. In the sections below, however, the most important thing is that you should follow the conceptual arguments.

Here is the situation we are trying to address. The mtcars data set contains $n = 19$ cars with automatic transmissions and $n = 13$ cars with manual transmissions. The $n = 19$ and $n = 13$ are **independent samples** that are standing in for the whole *population* of cars (from model year 1974). The samples are independent because they were collected from two distinctive groups of cars (as opposed to one group of cars at two different points in time). What we want to do is use the sample to infer a plausible difference in mileage between the two types of transmissions among all 1974 model year cars. I use the word *plausible* intentionally, as I want to avoid the language of probability until a little later. The key thing to keep in mind is that because we are using samples, we can't possibly know *exactly* what would be true of the populations. But we can think about the evidence that we get from the sample data to see if it convinces us that a difference may exist between the respective populations (i.e., the population of automatic cars and the population of manual cars). We will assume for the purposes of this example that the people involved in collecting the mtcars data did a good job at obtaining random samples of cars to include in the data set.

Of additional importance, we can use the statistics to give us some information about how far apart the fuel economy is between the two types of transmissions. We might have wanted to know this information to make a judgment as to which type of transmission to buy: automatics are more convenient and may be safer, but manuals might get better fuel economy and some people think they are more fun to drive. If the statistics told us that there was a big difference in fuel economy, we would have to weigh that against purchase and operating costs as well as convenience, safety, and fun. On the other hand, if the statistics told us that there was only a very small economy difference or possibly no difference between the two types of transmission, then we could make our decision based on other criteria.

Let's begin with some exploration of the mtcars data set, including a new visualization called a **box plot**. Run these commands:

```
mean( mtcars$mpg[ mtcars$am == 0 ] )     # Automatic transmissions
mean( mtcars$mpg[ mtcars$am == 1 ] )     # Manual transmissions
```

The mean miles per gallon (mpg) for the automatic transmission cars was 17.1, while the mean mpg for manual transmission cars was 24.3, a substantial difference of about 7.2 miles per gallon. Note that we have stepped up our game here with respect to R syntax. We are using the $ subsetting mechanism to access the mpg variable in the mtcars data set: that's what "mtcars$mpg" does. But we are also doing another kind of subsetting at the same time. The expressions inside the square brackets, [mtcars$am == 0] and [mtcars$am == 1] select subsets of the cases in the data set using logical expressions. The two equals signs together makes a logical test of equality, so for the first line of code we get every case in the mtcars data frame where it is true that mtcars$am is equal to 0 (0 is the code for automatic transmission; you can verify this by examining the help file you get when you type ?mtcars). For the second line of code we get every case where it is true that mtcars$am is equal to one (one is the code for manual transmission).

Now if you have not yet put on your brain-enhancing headgear, you might believe that the calculated difference in means is sufficient evidence to conclude that manual transmissions are better, as the mean for manual transmission cars is more than 7 mpg higher than for automatic transmission cars. Remember the previous chapter, however, and keep in mind how often we drew samples that were a fair bit different from the population mean. Each of these two sample means is uncertain: each mean is what statisticians refer to as a **point estimate**, with the emphasis on the word "estimate." Each sample mean is an estimate of the underlying population mean, but right now we are not quite sure how good an estimate it is. One of the key goals of inferential statistics is to put some boundaries around that level of uncertainty.

We can begin to understand that uncertainty by examining the variability within each of the groups. All else being equal, a sample with high variability makes us less certain about the true population mean than a sample with low variability. So as part of our routine process of understanding a data set, let's examine the standard deviations of each of the transmission groups:

```
sd( mtcars$mpg[ mtcars$am == 0 ] )     # Automatic transmissions
sd( mtcars$mpg[ mtcars$am == 1 ] )     # Manual transmissions
```

These commands reveal that the standard deviation for automatic transmissions cars is 3.8 miles per gallon, while the standard deviation for manual transmissions is quite a bit higher at 6.1 miles per gallon. Are these standard deviation values unexpected, very large, very small, or just what the doctor ordered? We really don't know yet, and in fact it is a little tricky to judge just on the basis of seeing the two means and the two standard deviations. We might get a better feel for the comparison between these two groups with a visualization that allows us to graphically compare distributions and variability. That's where the box plot comes in:

```
boxplot(mpg ~ am, data=mtcars)      # Boxplot of mpg, grouped by am
```

This command introduces another little piece of R syntax, called "formula notation." The expression "mpg ~ am" tells R to use mpg as the dependent variable (the variable that gets plotted on the Y-axis) and to group the results by the contents of "am." The second piece, "data=mtcars," simply tells R where to find the data that goes with the formula. Lots of analysis commands in R use the formula notation, and we will expand our knowledge of it later in the book. For now, take a look at the box plot that appears in Figure 4.1.

The box plot, sometimes also called a "box-and-whiskers plot," packs a lot of information into a small space. Figure 4.1 shows boxes and whiskers for the two groups of cars—automatic and manual—side-by-side. In each case the upper and lower boundaries of the box represent the first and third quartiles, respectively. So 25% of all the cases are above the box and 25% are below the box. The dark band in the middle of the box represents the median. You can see clearly that in the case of manual transmissions, the median is quite close to the first quartile, indicating that 25% of cases are clustering in that small region between about 21 and 23 miles per gallon. In this box plot the whiskers represent the position of the maximum and minimum values, respectively. In some other box plots these whiskers may represent the lowest or highest "extreme" value, with a few additional outliers marked beyond the whiskers. Other box plots may also notate the mean, sometimes with a dot or a plus sign.

Figure 4.1 gives us a good visual feel for the differences between the two groups. The boxes for the two groups do not overlap at all, a very intuitive and informal indication that there may be a meaningful difference between these

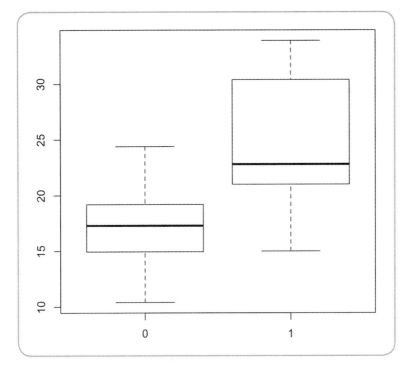

FIGURE 4.1. Box plot of mpg by transmission type from the mtcars data set.

two groups. In comparing the heights of the boxes, we also see a reflection of what we learned before from the standard deviations: the automatic transmission cars are considerably less variable than the manual transmission cars. Finally, the whiskers show that the very lowest value for manual transmissions falls at the first quartile for the automatic transmissions, further reaffirming the differences between these groups.

But we are still cautious, because we know that samples can fluctuate all over the place, and we can't be certain that the differences between these two groups can be trusted. So now let's do something clever: we can use the resampling and replication techniques we developed in the previous chapter to create a simulation.

EXPLORING THE VARIABILITY OF SAMPLE MEANS WITH REPETITIOUS SAMPLING

This simulation will show, in an informal way, the amount of uncertainty involved in these two samples. Let's try to visualize those boundaries using some of the tricks we learned in the previous chapter. First, we will have a little fun by sampling from our samples:

```
mean( sample(mtcars$mpg[ mtcars$am == 0 ],size=19,replace=TRUE) )
mean( sample(mtcars$mpg[ mtcars$am == 1 ],size=13,replace=TRUE) )
```

These functions should be familiar to you now: We are drawing a sample of $n = 19$ from the automatic transmission group, with replacement. Likewise, we are drawing a sample of $n = 13$ from the manual transmission group. I got 16.8 mpg for automatic and 28.4 mpg for manual, but your mileage will vary (ha ha!) because of the randomness in the sample() command. It may seem kind of goofy to "resample" from a sample, but bear with me: we are building up toward creating a histogram that will give us a graphical feel for the uncertainty that arises from sampling. Each of the sample mean differences will be close to, but probably not exactly equal to, the mean difference that we observed between the two original samples. Now, let's calculate the difference between those two means, which is, after all, what we are most interested in:

```
mean(sample(mtcars$mpg[mtcars$am == 0],size=19,replace=TRUE)) -
    mean(sample( mtcars$mpg[mtcars$am == 1],size=13,replace=TRUE) )
```

If you are typing that code, put the whole thing on one line and make sure to type the minus sign in between the two pieces. Keep in mind that we are usually looking at *negative* numbers here, because we are expecting that manual transmission mileage will on average be higher than automatic transmission mileage. Now let's take that same command and replicate it one hundred times:

```
meanDiffs <- replicate(100, mean( sample(mtcars$mpg[ mtcars$am == 0 ],
    size=19,replace=TRUE) ) - mean( sample(mtcars$mpg[ mtcars$am ==
    1 ], size=13,replace=TRUE) ))
```

Now plot a histogram of that sampling distribution of mean differences:

```
hist(meanDiffs)
```

The first statement above uses replicate() to run the same chunk of code 100 times, each time getting one sample mean from the automatic group and one sample mean from the manual group and subtracting the latter from the former. We store the list of 100 sample mean differences in a new vector called mean-Diff and then request a histogram of it. The result appears in Figure 4.2.

Let's make sense out of this histogram. The code says that whenever we draw a sample of automatic transmission cars, we calculate a mean for it. Then we draw a sample of manual transmission data and calculate a mean for it. Then we subtract the manual mean from the automatic mean to create a mean difference between two samples. Every time we do this we end up with a slightly different result because of the randomness of random sampling (as accomplished by the sample() command). We append each mean difference onto a vector of mean differences. As the histogram shows, in a lot of cases the difference between the two means is right around –7, as you would expect from looking at the two original samples. But we can also see that on occasion manual transmission samples are better by as much as 13 miles per gallon (the left end of the X-axis), while in other cases manual transmissions are only better by 1 mile per gallon (the right end of the X-axis). Note that if you run this code yourself, your results may be slightly different because of the inherent randomness involved in sampling.

You can think of this informally as what might have happened if we had replicated our study of transmissions and fuel economy 100 times. Based solely on the simulation represented by this histogram, we might feel comfortable saying this: manual transmissions may provide better fuel economy than automatic transmissions with a difference of about 7 miles per gallon, but that could on rare occasions be as little as 1 mile per gallon or as much as 13 miles per gallon. The width of the span between –1 and –13 is one very concrete representation of our uncertainty. Further, one might say that we have a certain amount of *confidence* that we do have a real difference between the two kinds of transmissions. I love that word "confidence." We are using it

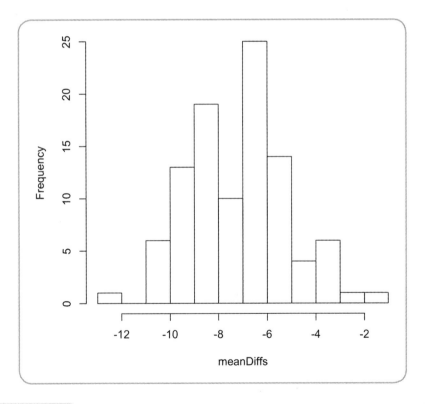

FIGURE 4.2. Histogram of mean differences between automatic transmission cars and manual transmission cars.

to signify that there is a span of different possibilities and although we know roughly how wide that span is, we don't know where exactly the truth lies within that span.

If we really wanted to follow the methods described in Chapter 3, we could go one step further by calculating quantiles for our distribution of mean differences. The following command provides the values for the 0.025 and 0.975 quantiles: this divides up the distribution of simulated sampling means into 95% in the center and 5% in the tails.

```
quantile(meanDiffs, c(0.025, 0.975))
```

For my simulated distribution of 100 mean differences, I got −10.8 on the low end and −3.1 on the high end. So we could now update our previous informal statement of uncertainty to become slightly more specific: manual transmissions may provide better fuel economy than automatic transmissions with a difference of about 7 mpg, but for 95% of the simulated mean differences that could be as much as 10.8 mpg or as little as 3.1 mpg. The width of this span, about plus or minus 4 mpg, is a representation of our uncertainty, showing what might happen in about 95 out of 100 trials if we repeatedly sampled fuel economy data for cars with the two types of transmissions.

OUR FIRST INFERENTIAL TEST: THE CONFIDENCE INTERVAL

Now we are ready to perform our very first official inferential test:

```
t.test(mtcars$mpg[mtcars$am==0] ,mtcars$mpg[mtcars$am==1])
```

That command produces the following output:

```
        Welch Two Sample t-test
data: mtcars$mpg[mtcars$am == 0] and mtcars$mpg[mtcars$am == 1]
t = -3.7671, df = 18.332, p-value = 0.001374
alternative hypothesis: true difference in means is not equal to 0
95 percent confidence interval:
-11.280194 -3.209684
sample estimates:
mean of x mean of y
17.14737 24.39231
```

The t.test() function above invokes the single most popular and basic form of inferential test in the world, called the "Student's *t*-Test." If you go all the way back to the introduction to this book, you will remember that "Student" was William Sealy Gosset (1876–1937), the Guinness Brewery scientist. Gosset developed this "independent groups" *t*-test in order to generalize to a population of mean differences using sample data from two independent groups of observations. The output on page 60 is designated the "Welch Two Sample *t*-test" because the 20th-century statistician Bernard Lewis Welch (1911–1989) developed an adaptation of Gosset's original work that made the test more capable with different kinds of "unruly" data (more specifically, situations where

Formulas for the Confidence Interval

I promised to postpone a deeper discussion of the meaning of *t* until the next chapter, but now is the right moment to show you the formulas for the confidence interval for the difference between two independent means:

Confidence interval: Lower bound $= \left(\overline{x}_1 - \overline{x}_2\right) - t^{\star}\sqrt{\dfrac{s_1^2}{n_1} + \dfrac{s_2^2}{n_2}}$

$$\text{Upper bound} = \left(\overline{x}_1 - \overline{x}_2\right) + t^{\star}\sqrt{\frac{s_1^2}{n_1} + \frac{s_2^2}{n_2}}$$

You'll notice that the top and bottom equations only have one difference between them: the top equation has a minus sign between the first and second part and the bottom equation has a plus sign. The first half of each equation, a subtraction between two "*x*-bars," is simply the observed difference in sample means. In our mtcars example, that difference was $17.14 - 24.39 = -7.2$. The second part of the equation calculates the width of the confidence interval, in the top case subtracting it from the mean difference and in the bottom case adding it.

The width of the confidence interval starts with t^*—this is a so-called critical value from the *t*-distribution. I won't lay the details on you until the next chapter, but this critical value is calculated based on the sample sizes of the two samples. The important thing to note is that the critical value of *t* will differ based on both sample size and the selected confidence level. We have used a "95% confidence interval" throughout this chapter, but it is also possible to use 99% or on occasion other values as well.

All of the stuff under the square root symbol is a combination of the variability information from each of the samples: technically a quantity called the **standard error.** Sounds complicated, but it is really nothing more than the standard deviation of the sampling distribution of means (or in this case, mean differences). In each case we square the standard deviation to get the variance and then divide the variance by the sample size. Once we have added together the two pieces, we square root the result to get the standard error.

the two groups have different levels of variability; Welch, 1947). We are going to postpone a detailed consideration of what "*t*" actually means until the next chapter. So for now, the key piece of output to examine above is the **95 percent confidence interval.** The *t*-test procedure has used these two samples to calculate a confidence interval ranging from a mean difference of −11.3 miles per gallon to −3.2 miles per gallon. That range should seem familiar: it is darned close to what our little resampling simulation produced! In fact, there is some conceptual similarity to what we did in our informal simulation and the meaning of a confidence interval.

In our simulation we sampled from the existing sample, because we had no way of sampling new data from the population. But statisticians have figured out what would happen if we *could* have sampled new data from the population. Specifically, if we *reran our whole study of transmissions and fuel economy many times*—sampling from the population and taking means of both a new group of automatic transmission cars and a new group of manual transmission cars—and each time we constructed a new confidence interval, in 95% of those replications the confidence interval *would contain the true population mean difference.* In the previous sentence the phrase "would contain" signifies that the true population mean difference would fall somewhere in between the low boundary of the confidence interval and the high boundary. Based on this definition it is really, extremely, super important to note that *this particular confidence interval* (the one that came out of our *t*-test above) does not necessarily contain the true population value of the mean difference. Likewise, the 95% is *not* a statement about the probability that *this particular confidence interval* is correct. Instead, the 95% is a long-run prediction about what would happen if we replicated the study—sampling again and again from the populations—and in each case calculated new confidence intervals.

This is definitely a major brain stretcher, so here's a scenario to help you think about it. You know how in soccer, the goal posts are fixed but the player kicks the ball differently each time? Now completely reverse that idea in your mind: pretend that the player does the same kick every time, but you get the job of moving the goal posts to different locations. In fact, let's say you get 100 tries at moving the goal posts around. A 95% confidence interval indicates that 95 out of 100 times, you moved the goal posts to a spot where the mystery kick went right through. The player always does the same kick: that is the unseen and mysterious population mean value that we can never exactly know. Each of the 100 times that you move the goal posts represents a new experiment and a new sample where the two posts are the two edges of the confidence interval calculated from that sample. You can create a nifty animation in R that helps to demonstrate this idea with the following code:

```
install.packages("animation")
library(animation)
conf.int(level=0.95)
```

The animation creates 50 different confidence intervals around a population mean of 0 by repeatedly sampling from the normal distribution. The dotted line in the middle of the graph shows where the population mean lies. Each
constructed confidence interval looks like a tall capital letter "I." The circle in
the middle of the "I" shows where the mean of each sample falls. In a few cases,
the confidence interval will not overlap the population mean: these are marked
in red. Most of the time these "goal posts" overlap the population mean, but in
about 5% of the cases they do not. If for any reason this code did not work for
you in R, try to search online for "confidence interval simulation" and you will
find many animations you can view in a browser.

Now back to the mtcars transmission data: remember that when we say
"95% confidence interval," we are referring to the proportion of constructed
confidence intervals that would likely contain the true population value. So
if we ran our transmission and fuel economy study 100 times, in about 95 of
those replications the samples of transmission data would lead to the calculation of a confidence interval that overlapped the true mean difference in
mpg. As well, about five of those 100 replications would give us a confidence
interval that was either too high or too low—both ends of the confidence
interval would either be above or below the population mean (just like the
red-colored intervals in the animation described above). And for the typical
situation where we only get the chance to do one study and draw one sample,
we will never know if our particular confidence interval is one of the 95 or
one of the five.

From the *t*-test, the span of −11.3 up to −3.2 is what statisticians call an
interval estimate of the population value. The fact that it is a range of values
and not just a single value helps to represent uncertainty: we do not know and
can never know exactly what the population value is. The width of the confidence interval is important! A wide interval would suggest that there is quite a
large span where the population mean difference may lie: in such cases we have
high uncertainty about the population value. A narrow interval would signify
that we have a pretty sharp idea of where the population mean difference is
located: low uncertainty. We would want to keep that uncertainty in mind as
we think about which kind of automobile transmission to choose, based on our
preferences about fuel economy as well as other criteria such as cost.

Finally, I want to emphasize again the importance of the idea that the
observed confidence interval does not tell us where the true population mean
difference lies. The population mean difference *does not* lie at the center point of
the confidence interval. The confidence interval we calculate from our sample
of data may, in fact, not even contain the actual population mean difference. In
keeping with our consideration of inferential thinking, the confidence interval
adds to the weight of evidence about our beliefs. The span of −11.3 up to −3.2
strengthens the weight of evidence that the population difference in fuel economy between automatic and manual transmissions is a negative number somewhere in the region of −7.2 mpg plus or minus about 4 mpg. The confidence

interval does not *prove* that there is a difference in fuel economy between the two types of transmissions, but it does suggest that possibility, and it gives us a sense of the uncertainty of that conclusion (the plus or minus 4 mpg represents the uncertainty).

CONCLUSION

We began this chapter by constructing our own informal simulation model of a sampling distribution of mean differences by extrapolating from two samples of data about the fuel economy of 1974 cars. This was only an informal model because we did not have access to the complete populations of car data, but the simulation did give us the chance to put the idea of the uncertainty into a graphical context.

We then conducted a *t*-test, which calculated a confidence interval based upon the two samples of car data. The confidence interval suggested that manual transmissions from the 1974 model year might be more efficient than automatic transmissions, by somewhere in the neighborhood of 7 mpg. The interval estimate of the population mean difference ranged from −11.3 up to −3.2, a span of about 8 mpg with the observed mean difference between the samples −7.2, right in the center of that range. The width of the confidence interval, that is, the plus or minus 4 from the center point of −7.2, was an indication of our uncertainty. If the interval had been wider we would have been less certain. If the interval had been narrower, we would have been more certain.

Throughout this process, we firmly held in mind the idea that the meaning of a 95% confidence interval is that in 95 out of 100 study replications, we would be likely to find that whatever confidence interval we constructed in a given study did contain the actual population value—in this case a mean difference in fuel economy between two types of cars. Similar to the concepts we explored in the previous chapter, this is a statement about probabilities over the long run and not a statement about the particular confidence interval we constructed from this specific data set. This particular confidence interval may or may not contain the true population value: we will never know for sure.

EXERCISES

1. In your own words, write a definition of a 95% confidence interval.

2. Answer the following true/false questions about confidence intervals: (a) The center of the confidence interval is the population value; (b) The confidence interval always contains the population value; (c) A wider confidence interval is better, because it signals more certainty; (d) When we say "95% confidence interval," what we mean is that we are 95% certain that this particular confidence interval contains the population value.

3. Run the code shown in this chapter that created the animation of confidence intervals:

```
install.packages("animation")
library(animation)
conf.int(level=0.95)
```

Once the animation has finished running, comment on your results. Pay particular attention to the number of times that the confidence interval did not contain the population mean value (0). You may have gotten a different answer from other people who completed this exercise. Explain why this is so in your own words.

4. Some doctors conducted clinical trials on each of two new pain relievers. In the first trial, Drug A was compared to a placebo. In the second trial, Drug B was also compared to a placebo. In both trials, patients rated their pain relief such that a more negative number, such as −10, signified better pain relief than a less negative number, such as −5. As you may have already guessed, a rating of 0 meant that the patient's pain did not change, and a positive rating meant that pain actually increased after taking the drug (yikes!). After running the trials, the doctors calculated confidence intervals. Drug A had a confidence interval from −10 to −2 (these are mean differences from the placebo condition). Drug B had a confidence interval from −4 to +2 (again, mean differences from the placebo condition). Which drug is better at providing pain relief and why? Which drug gives us more certainty about the result and how do you know?

5. Assume the same conditions as for the previous question, but consider two new drugs, X and Y. When comparing Drug X to placebo, the confidence interval was −15 to +5. When comparing Drug Y to placebo, the confidence interval was −7 to −3. Which drug is better at providing pain relief and why? Which drug gives us more certainty about the result and how do you know?

6. Use the set.seed() command with the value of 5 to control randomization, and then calculate a confidence interval using the rnorm() command to generate two samples, like this:

```
set.seed(5)
t.test(rnorm(20,mean=100,sd=10),rnorm(20,mean=100,sd=10))
```

The set.seed() function controls the sequencing of random numbers in R to help with the reproducibility of code that contains random elements. Review and interpret the confidence interval output from that t.test() command. Keep in mind that the two rnorm() commands that generated the data were identical and therefore each lead to the creation of a sample representing a population with a mean of 100. Explain in your own words why the resulting confidence interval is or is not surprising.

7. The built-in PlantGrowth data set contains three different groups, each representing a different plant food diet (you may need to type data(PlantGrowth) to activate it). The group labeled "ctrl" is the control group, while the other two groups are each a different type of experimental treatment. Run the summary() command on PlantGrowth and explain the output. Create a histogram of the ctrl group. As a hint about R syntax, here is one way that you can access the ctrl group data:

PlantGrowth$weight[PlantGrowth$group=="ctrl"]

Also create histograms of the trt1 and trt2 groups. What can you say about the differences in the groups by looking at the histograms?

8. Create a boxplot of the plant growth data, using the model "weight ~ group." What can you say about the differences in the groups by looking at the boxplots for the different groups?

9. Run a *t*-test to compare the means of ctrl and trt1 in the PlantGrowth data. Report and interpret the confidence interval. Make sure to include a carefully worded statement about what the confidence interval implies with respect to the population mean difference between the ctrl and trt1 groups.

10. Run a *t*-test to compare the means of ctrl and trt2 in the PlantGrowth data. Report and interpret the confidence interval.

CHAPTER 5

Bayesian and Traditional Hypothesis Testing

Let's pick up where we left off with the mtcars data set and our comparison of types of transmissions in a bunch of 1974 model year cars. Here's an interesting fact about the *t*-test: when William Sealy Gosset thought about the data he collected about two different experimental groups of beer, he began with the assumption that a control group represented an **untreated** population, and, somewhat pessimistically, that the **treatment** population would be essentially the same as the control group. In other words, his reasoning *began with the idea that the mean difference between the two groups would be zero* and it was up to the sample data to provide any evidence to the contrary. In our mtcars example, we could think about the manual transmission group as the control group and the automatic transmission group as the treatment group. (It is a somewhat arbitrary choice in this scenario; automatic transmissions were invented after manual transmissions, if that helps.) Gosset would have started with the assumption that buying a car with an automatic transmission would do nothing to change the fuel economy in comparison with the manual transmission control group. (By the way, the automatic transmission was invented in 1904, and Gosset lived until 1937, so he may have actually ridden in a car with an automatic transmission.)

Bayesian thinking takes a different tactic by sidestepping the assumption that the two populations are the same and allowing us to assert a prior belief that the two types of transmissions may differ. While the control group (manual transmission) still provides a baseline, the treatment group (automatic transmission) provides new information that informs an improved understanding of the differences between the two groups. In other words, new observations from the automatic transmission group give us a clearer and clearer perspective on how different the two groups may be. This idea lies at the heart of Bayesian thinking: each new piece of evidence updates and modifies what we previously believed to be true. Using this method, instead of just seeing an upper and a lower bound to the mean difference, we can actually see the probabilities of various amounts of difference between the treatment and control groups.

Before we dig in more deeply, it may be helpful to consider the origins of Bayesian thinking so that we know what to look for as we examine our data and analytical results. The Bayes moniker comes from Thomas Bayes (1702–1761), a Presbyterian minister with a lifelong interest in mathematics. His most famous paper, and the one that gives us Bayes' theorem, was actually published a couple of years after his death (Bellhouse, 2004). Put simply, Bayes' theorem indicates that we can estimate the probability of a particular scenario in light of new evidence if we have some other probability information. We have already accomplished this ourselves with our toast-drop example from Chapter 2, hereby reproduced for your viewing convenience as Table 5.1.

Bayes' theorem gives us the tools to answer questions like this: There's a piece of toast lying on the ground with the topping facing down, what's the probability that the topping is jelly? In Bayesian terms, we want to know p(jelly|down). You can say this mathematical phrase as the "probability of observing jelly, given that the toast is facing down." To figure this out with Bayes' theorem, we need to know p(jelly), which is 0.3, and p(down), which is 0.5. These are the so-called base rates—in Table 5.1 the marginal probabilities of having jelly and toast-facing-down. So p(jelly) is a row total and p(down) is a column total. Finally, we need to know p(down|jelly)—that is, the probability of finding a piece of toast face down, given that it is topped with jelly. In Table 5.1, p(down|jelly) is 0.2/0.3 = 0.667 because in the top row, two out of a total of three of the jelly scenarios have the toast face down. Put this altogether according to Bayes and you have the following equation: p(jelly|down) = (0.667 × 0.3)/0.5 = 0.40, also known as 40%.

I actually liked the methods that we used in Chapter 2 better than this formula, but the two strategies are mathematically and functionally equivalent. If we apply the same Bayesian thinking to our mtcars example, our question might become this: What's the probability of having a difference of greater than 3 mpg in fuel economy between manual and automatic transmissions, given the sample data we have collected? I chose 3 mpg because it happened to be the bottom end of the confidence interval; you could pick any value you wanted. Think of our greater-than-3 mpg scenario as one of the cells in the contingency

TABLE 5.1. Two-by-Two Contingency Table for Bayesian Thinking			
	Down	**Up**	**Row totals**
Jelly	.2	.1	.3
Butter	.3	.4	.7
Column totals	.5	.5	1.0

Note. This table is the same as Table 2.8.

Notation, Formula, and Notes on Bayes' Theorem

Bayes' theorem is usually illustrated with an equation that looks something like this:

$$p(H \mid D) = \frac{p(D \mid H)\, p(H)}{p(D)}$$

Given that this is a book about statistical inference and hypothesis testing, I have used H to stand for hypothesis and D to stand for data. Let's dissect this equation piece by piece. What we are trying to find out is $p(H|D)$, the probability of this hypothesis being "true" given the statistical outcomes we observed from these data. In Bayesian terms this is known as the **posterior probability**. We need three ingredients in order to figure this out:

- The probability of observing these data when the hypothesis is "true," $p(D|H)$, which Bayesian folks call the **likelihood**;
- The **prior probability,** $p(H)$, in other words our baseline belief about the "truth" of the hypothesis;
- The probability of observing these data under any and all conditions, $p(D)$, or in Bayes-speak, the **evidence.**

Now we can restate our mathematical equation with the Bayesian words filled in:

$$\text{posterior} = \frac{\text{likelihood} \cdot \text{prior}}{\text{evidence}} = \text{prior} \left(\frac{\text{likelihood}}{\text{evidence}} \right)$$

Conceptually a lot of people like to think about this equation as a way of converting *prior* beliefs into *posterior* probabilities using new data. The new data shows up twice, once in *likelihood* and once in *evidence*. In fact, the ratio of likelihood/evidence is just a way of normalizing expectations about these data based on their marginal probability (see Table 5.1).

table as indicated in Table 5.2. In the rows, we have different beliefs about whether the type of a car's transmission makes any difference to fuel economy. In the columns we have the observations we might obtain from samples, in some cases a mean difference of greater than 3 mpg, and in other cases not. The cells and marginal totals are blank at this point, but you could figure out ways of filling them in. For the row totals, we could ask some automotive experts about their beliefs. For the column totals we could repeatedly resample from our data to find out what proportion of the mean differences were greater than 3 mpg. The most important conceptual takeaway is that Bayes' theorem gives

TABLE 5.2. Bayesian Thinking about Mean Differences			
	> 3 MPG	< 3 MPG	Row totals
Transmissions do matter			
Transmissions don't matter			
Column totals			

us a way of reasoning about a hypothesis in the light of evidence we collect in a sample. Bayes' theorem transforms our prior beliefs into posterior probabilities using evidence from new data.

Thanks to the great community of statisticians who are always at work on improving R, there are several new packages available to put this Bayesian thinking into practice. In the case of the *t*-test of two independent samples (which is what we ran in Chapter 4 to get the confidence interval), one great place to start is the BEST package, which stands for Bayesian Estimation Supersedes the *t*-test (Kruschke, 2013; Meredith & Kruschke, 2015). Later in the book, we will use other packages for the Bayesian approach, but the BEST package stands out for its powerful calculation methods and its excellent and informative graphics.

Note one minor complication with installing the BEST package. Most R packages are fully self-contained, in that they contain all of the code they need or they refer to other R packages that provide what they need. For the BEST package, we also need a small piece of extra software called JAGS (Just Another Gibbs Sampler; see *mcmc-jags.sourceforge.net*). JAGS is free, open-source software that runs on Windows and Mac, but you must install it separately and in advance of asking R to install the BEST package. If you have difficulty, try searching for "install JAGS for R," and you will find some useful resources. If all else fails, statistician Rasmus Baath created a web page where you can run the same analysis described below in a browser without using R (see *www.sumsar.net/ best_online*).

The BEST package runs a simulation called "Markov chain Monte Carlo" in order to characterize the distances between the means of the two groups. A *Markov chain* is a statistical technique for examining the probabilities of linked events. "Monte Carlo" refers to a process of running repeated simulations in order to look around for everything in between the "best case" and "worst case" scenarios. I ran the following code to get results from the BEST package:

```
install.packages("BEST")
library(BEST)
carsBest <- BESTmcmc(mtcars$mpg[mtcars$am==0],
    mtcars$mpg[mtcars$am==1])
plot(carsBest)
```

Markov-Chain Monte Carlo Overview

The Markov-Chain Monte Carlo, commonly abbreviated as MCMC, combines two things that statisticians love to do: sample from distributions and evaluate solutions. Sampling from distributions is a piece of cake: we did lots of sampling in Chapter 3. For example, right at the beginning of the chapter, we sampled from the uniform distribution with the runif() command. Monte Carlo statistical techniques, at the most basic level, involve repeatedly sampling from a known distribution (i.e., the parameters, like mean and standard deviation, are set), computing some output from each sample (e.g., an estimate of the population mean), and compiling these results in order to reason about the estimation process.

A Markov chain, named after the Russian mathematician Andrey Markov (1856–1922), is a mathematical structure representing a matrix of state changes with probabilities. Imagine a king making a move in a chess game: the king may move to any of eight neighboring squares, but nowhere else. The chances of moving to a particular neighboring square depend on whether the square is occupied by another piece or if the move would put the king in danger. Markov chains model processes where the possibilities for the next move only depend on the current position.

When conducting a Bayesian analysis, a mathematical approach to figuring out the likelihood, prior, and evidence (see "Notation, Formula, and Notes on Bayes' Theorem") often becomes too complex to be solved with a single computation, so we must use some kind of repetitive analysis process to find a solution. Put Monte Carlo and Markov chain together and you get a powerful combination for sampling, evaluating, and improving. We begin with an arbitrary candidate for the parameter we are modeling. We evaluate the posterior probability in light of our data. Then we change the parameter a little bit and evaluate the posterior probability again. We compare these two probabilities using a method of evaluation (e.g, a technique called Metropolis–Hastings) and decide whether to keep the parameter change based, in part, on whether the new parameter seems more probable. If we keep the parameter, that becomes our new "location" in the Markov chain. By repeating this process thousands of times we adjust the parameter in a series of random moves that gradually builds up a whole distribution of likely parameter values. One important result of this process is that rather than getting a single "point estimate" of a parameter, or even the two ends of a confidence interval, we end up with an actual distribution of a range of parameter values (more often than not, this looks something like the normal distribution). We can examine this actual distribution, using histograms and descriptive statistics, and see which ranges of parameter values are more likely and which are less likely.

If that all seems a little abstract, you should try a wonderful app called MCMC Robot, developed by biologist Paul O. Lewis. Search online for MCMC Robot and you will find a Windows version and an iOS version (for iPad). MCMC Robot uses

(continued)

a simple concept: A hypothetical robot walks around on a landscape that you create in order to find the highest ground (i.e., the region surrounding, near, and at the top of the highest hill). The population parameter we are modeling represents the altitude of the fictitious landscape. The path of the robot appears on the screen as a series of dots connected by lines. Each new step that the robot takes—left, right, forward, or back—is random and only related to where it stands now, but steps that seem to lead upward are more highly favored than those that stay at the same altitude or that seem to lead downward. These dots and lines represent the Markov chain. When the robot walks for thousands of steps, you can see clearly how the path of the robot gradually heads toward higher ground and ends up walking around randomly near the top of the hill. By compiling the altitudes of all of the steps (leaving out the first few "burn in" steps while the robot locates the "foothills"), one obtains a distribution of the population parameter. Very cool!

The install.packages() function downloads the code from a repository on the Internet that the creators of R have set up, while the library() function brings the new package into the computer's memory so that we can use it. The BESTmcmc() function computes a probability distribution for the mean differences between the two groups, using the full information available in the two samples of transmission data. The final function above is a call to plot(), which has been enhanced by the BEST package with the ability to show the specialized output from the BESTmcmc() function. The result appears in Figure 5.1.

The bell-shaped histogram in Figure 5.1 has several excellent features. First, the most highly likely mean difference value is noted at the top of the curve as −7.21 (mpg). Remember that this is quite similar, but not exactly identical, to the observed mean difference between the two samples. Given the steep dropoff of the probability curve (to the right) down to a tiny tail at about −2, it is also clear that the probability of no difference between transmission types is very small. This is exceedingly important for us to know as it clearly shows that manual transmissions in 1974 model year cars almost certainly had some degree of superiority in fuel economy over automatics.

Look on the right side of Figure 5.1 in the middle. There is an expression there that says 99.8% < 0 < 0.2%. This expression shows the proportion of mean differences in the MCMC run that were negative versus the proportion that were positive. Again here the evidence is unequivocal: 99.8% of the mean differences in the distribution were negative (meaning that manual transmissions were more efficient) and 0.2% were positive. This confirms what we already viewed in the distribution plot: the chances that 1974 automatic transmissions were equal to or better than manual transmissions are close to zero.

There is some new notation on this graph at the bottom. Those funny shaped things that look like the letter *u* are actually the Greek letter "mu."

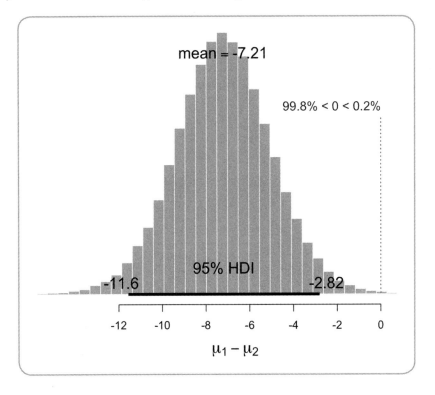

FIGURE 5.1. Bayesian probability distribution of the differences between two means.

Statisticians commonly use the mu symbol to indicate the population mean. So the expression at the bottom signifies "population mean of group 1 minus population mean of group 2." Conceptually, this is a meaningful and important idea. The Bayesian approach used in BESTmcmc() is stating that it has used the sample data for automatic transmissions and the sample data for manual transmission to provide a model of the difference in *population* means.

Finally, and perhaps most informatively, near the X-axis the graph shows the so-called **highest density interval,** or HDI, as being between −11.6 and −2.82 (mpg). This HDI has a simple and intuitive interpretation: there is a 95% probability that the population mean difference between the two groups falls within this range. The Bayesian reasoning that went into the construction of the BESTmcmc procedure allows us to make inferences about the population of mean differences based on the sample data we have analyzed. According to this reasoning, 95% of the likely values of the population mean difference lie in the bell-shaped area between −11.6 mpg and −2.82 mpg. Looking at the shape of the curve, it is clear that the greatest likelihood is for a population mean difference somewhere near −7.21—say roughly speaking in the region between −9 and −5. Other population mean differences are possible, of course, just not

Detailed Output from BESTmcmc()

In addition to the graph presented in Figure 4.2, it can be illuminating to review the detailed numeric output from the BESTmcmc() procedure. Recall that we directed the output of the procedure into a data object called carsBest. We can review more details by typing the name of that data object at the command line:

carsBest

For our comparison of automatic and manual transmissions, this produces the following output:

MCMC fit results for BEST analysis:
100002 simulations saved.

	mean	sd	median	HDIlo	HDIup	Rhat	n.eff
mu1	17.146	0.971	17.140	15.201	19.045	1.000	60816
mu2	24.361	1.994	24.365	20.358	28.274	1.000	58435
nu	37.527	29.967	29.195	2.126	97.566	1.001	24943
sigma1	3.992	0.7807	3.889	2.620	5.563	1.000	41165
sigma2	6.788	1.6687	6.520	4.061	10.133	1.000	34370

'HDIlo' and 'HDIup' are the limits of a 95% HDI credible interval.
'Rhat' is the potential scale reduction factor (at convergence, Rhat=1).
'n.eff' is a crude measure of effective sample size.

The beginning of the output shows that 100,002 steps occurred in the Markov chain Monte Carlo (MCMC) simulation. Each of those 100,002 steps generated one set of credible estimates of our population parameters. See "Markov chain Monte Carlo Overview" for a more detailed discussion of how the MCMC process accomplishes this. We then have a table of statistics describing each distribution of a population parameter, represented by the column headings: mean, standard deviation (SD), median, the lower bound of a highest density interval (HDI), the upper bound for an HDI, something called Rhat and something called n.eff. On the left edge we have the row headings, each of which represents a population parameter for which we are seeking the posterior distribution.

So, for instance, the number in the upper left of this table, 17.146, is a point estimate of the mean population value of the fuel efficiency of 1974 cars with automatic transmissions. The next four numbers in that line—SD (0.971), median (17.14), HDLlo (15.201), and HDLhi (19.045)—describe the distribution of the 100,002 estimates in the posterior distribution. So 95% of the estimates of the population mean for the automatic transmission group fell between 15.2 and 19.0 with the median right in the middle of these two at 17.14. We have parallel results showing the posterior distribution of the population mean of the manual transmission group.

(continued)

By the way, if you remember the histogram of mean differences that appeared in Figure 4.2, the way the data for that histogram was generated was simply by subtracting the manual transmission mean from the automatic transmission mean for each of the 100,002 steps in the MCMC process.

The distributions for the population standard deviation for each group appear in the last two lines of the table. You may want to compare these results to the original standard deviations in the mtcars raw data (*SD* = 3.83 for automatic; *SD* = 6.71 for manual). If you are finding it tricky to imagine the posterior distributions just from these descriptive statistics, you can produce histograms for all of the posterior distributions with the following command:

```
plotAll(carsBest)
```

The first four histograms along the left column of the plot output show the posterior parameter histograms for the two population means and the two population standard deviations.

The output table shown above also contains some diagnostic information that you may be curious about. In the middle of the table is a line labeled "nu" (a Greek letter), that is an indicator of the effective shape of each posterior distribution of means (Kruschke, 2013). Higher values—as a rule of thumb any values larger than 30—indicate a roughly normal distribution, while lower values indicate distributions with taller tails (and therefore more uncertainty). There is also a column for Rhat, a diagnostic value which, when close to 1.0, shows that the MCMC process "converged" properly on sensible results. Values larger than 1.1 might indicate a problem with the MCMC process that could be addressed by running additional steps. Likewise, n.eff stands for "effective sample size." Even though the overall MCMC process took 100,002 steps, some of those steps exhibited an unwanted phenomenon called autocorrelation, so we don't count them toward the effective sample size. If n.eff were ever to drop below 10,000 for one or more of the lines in the table, we would also want to add more steps to the MCMC process.

as likely as those in the central region near –7.21. Putting this altogether into a single sentence, we might say that the population mean difference is somewhere near –7.21 miles per gallon, with the 95% highest density interval ranging from –11.6 mpg to –2.82 mpg. We might add that the likelihood of a population mean difference of 0 or larger is 0.2%.

The HDI is a somewhat similar idea to the notion of the confidence interval from earlier in this chapter, but there are some important conceptual differences. The confidence interval is calculated directly from the sample data and conceptually provides no direct evidence about the population mean difference. Remember that the definition of a 95% confidence interval is that if we ran an experiment a large number of times (each time generating a new sample), then about 95% of the confidence intervals we constructed from those repetitions would actually contain the population mean difference. This is a statement about confidence intervals over the long run, not the particular one that we calculated. In contrast, the HDI is built up gradually from more than 100,000 steps in our Markov chain Monte Carlo process, with each step depicting a possible combination of the population parameters. A key distinction to reinforce: the BEST HDI directly models the population parameters of interest and shows us probability distributions for those parameters, whereas the confidence interval uses sample data to compute one and only one example of an upper and lower bound for the population mean.

Both the end points of the confidence interval and the end points of the HDI show a range of values, but the HDI offers more detail. In both cases the width of the interval does represent our uncertainty about the population mean difference, but because the HDI is presented in the context of a probability distribution of estimated population mean differences, we get a clearer picture of that uncertainty from the Bayesian analysis. We can see the extent of the tails that extend beyond the HDI and we can see the density of the central region that it contains. With the confidence interval, we only know the end points of one plausible interval.

Finally, the confidence interval we calculated assumes that a t distribution makes a good model of differences between population means estimated from two independent samples. The t distribution is a family of symmetric distributions similar to the normal distribution but with "thicker" tails that are a function of the greater uncertainty involved in working with samples. For two good-sized, normal-looking samples (say, each greater than 30 observations) with equal variances, the t distribution generally does a nice job as a model of population mean differences. For badly behaved samples, with skewness, unequal variances, or other problems such as outliers, the t distribution may not always be such a great model (Micceri, 1989; Nguyen et al., 2016; Sawilowsky & Blair, 1992). The BEST procedure, through the power of the MCMC process, makes no special assumptions about the shape of the posterior distribution. Distributional anomalies in the samples used in the analysis will tend to stretch the width of the HDI and raise the tails appropriately to signal the greater uncertainty in those samples.

Now we are finally ready to answer the original research question that we posed back in Chapter 4: Do cars with automatic transmissions get better or worse mileage than cars with manual transmissions? We are in a good position to judge that with the evidence we have gathered from our confidence interval (using the t-test) and from our HDI (from the Bayesian BEST procedure). Both strategies converged on the idea that the manual transmissions from 1974 cars provided better fuel economy than cars from that year with automatic transmissions, by somewhere around 7 mpg. The confidence interval helped us to quantify our uncertainty at plus or minus 4 mpg. The BEST HDI gave us highly detailed information about the probability distribution of the population mean difference between these two groups. If we were selling cars back in 1974, we would certainly have wanted to consult with customers and business experts to find out how to balance the convenience of the automatic with the worse fuel economy provided by that type of transmission. Surprisingly, I was not quite old enough to drive in 1974, but given a choice I would have opted for the manual transmission in order to get the better fuel economy.

This brief discussion also suggests how important it is to entwine the statistical evidence with the real research situation and in this case some human values about the usefulness of a particular technology. The confidence interval and the HDI give us the basic tools to understand the magnitude of difference we might expect, but we have to look at that magnitude in light of the real-world situation of cost, convenience, and enjoyment.

THE NULL HYPOTHESIS SIGNIFICANCE TEST

Most scientific publications accept the use of confidence intervals in scientific reports and many of the reviewers and editors of these publications appreciate their value. Likewise, Bayesian methods of statistical inference are gradually gaining in popularity. Yet through most of the 20th century, a different perspective on statistical inference predominated. Some people refer to this as the *frequentist* perspective, but the most important statistical term to grasp is the **null hypothesis significance test.** You may recall the idea that William Sealy Gosset considered a mean difference of zero to be the one to beat using a comparison of two sample means. A contemporary of Gosset's, Ronald Fisher (1890–1962), took this thinking a step further and developed a step-by-step procedure called the "null hypothesis significance test" (NHST) to make an easy step-by-step method for researchers to draw conclusions about sample data. Here are the essential steps in the NHST:

• Begin by asserting a null hypothesis that there is no mean difference between the means of two groups.

• Choose an "alpha level" probability, beyond which the null hypothesis will be rejected: a common alpha level is .05; a more stringent level could be .01 or .005.

• Collect data and conduct a statistical test, such as the *t*-test of two independent means, to calculate a significance value, designated by the letter *p*.

• If the calculated value of *p* is *less than the alpha level* that was chosen above—for example, if $p = .049$ when alpha was chosen as .05, then *reject* the null hypothesis.

• When the null hypothesis is rejected, this can be considered evidence in favor of some unspecified alternative hypothesis, though the results of that significance test do not say anything specific about what that alternative hypothesis might be, or the probability that any particular alternative hypothesis may be correct.

- If p is *greater than the alpha level* that was chosen above—for example, if p = .051 when alpha was chosen as .05, then *fail to reject* the null hypothesis. Failing to reject the null hypothesis does not mean that we accept the null hypothesis, rather that we have no good evidence either way. Likewise, the p-value does not inform the question of how likely the null hypothesis is.

NHST has been used in thousands of scientific articles and in many other research situations. It is still used today even though there have been years and years of concerns expressed about it. Haller and Krauss (2002) found that 88% of the scientists they surveyed could not reliably identify the correct logic for the NHST. The governing board of the American Statistical Association has responded to a host of criticisms of the NHST by publishing its own cautionary guidelines about significance testing (Wasserstein & Lazar, 2016). This method is still in use mainly because most researchers received their training in statistics at a time when NHST was the predominant method of statistical inference. As the newer strategies are taught more commonly, it is likely that the use of the NHST will diminish.

Even though the NHST has shown flaws, as used in practice, I believe it is important for you to understand the logic of it, as well as some of the criticisms that have been leveled against it. Without delving too deeply into the mathematical roots of the criticisms, I will discuss three points that many people use when they are criticizing the NHST. Before doing that, however, I want to excerpt some material from the t-test output that I presented in Chapter 4:

```
t.test(mtcars$mpg[mtcars$am==0], mtcars$mpg[mtcars$am==1])
Welch Two Sample t-test
data: mtcars$mpg[mtcars$am == 0] and mtcars$mpg[mtcars$am == 1]
t = -3.7671, df = 18.332, p-value = 0.001374
alternative hypothesis: true difference in means is not equal to 0
sample estimates:
mean of x mean of y
17.14737 24.39231
```

I have left out the little chunk of the output that focused on the confidence intervals and I have retained the portion that contains information pertinent to the NHST. First, let's assume that we chose an alpha level of .05, even though we have not yet discussed what that means. In general, choosing an alpha of .05 is the most common choice that appears in the research literature, particularly among social scientists. In reference to this alpha, the p level shown above, p = .001374, is notably smaller than the alpha threshold. Thus, based on the conventional rules of the NHST, we reject the null hypothesis. Although this is not direct support for any particular alternative hypothesis, the output from the procedure notes that the "alternative hypothesis: true difference in means is not equal to 0." This is the logical opposite of the null hypothesis, so by rejecting the null hypothesis, we do express support for the logical opposite, specifically in this case that the difference in fuel economy between automatic and manual transmissions in 1974 model year cars is not equal to zero.

In a typical 20th-century research article, the results noted above—that we rejected the null hypothesis—would be cause for major celebration including party hats and cake. The fact that no other information is provided, for example, about the magnitude of the likely difference between the means, is absent from many published articles. This is the first criticism of the NHST, that by providing a simple go/no-go decision on research results, the results of the method have limited meaning. In the strictest interpretation of these NHST results, the statistician would go back to the engineers and the business experts and say, "Yes, by rejecting the null hypothesis we can firmly say that the difference between the two transmissions is not zero." We know, however, that the magnitude (and the direction) of difference in fuel economy between the two types of transmissions is very important. Having tests that simply tell us whether a difference is statistically significant is not especially useful on its own.

One way that statisticians have helped researchers to cope with this problem is by developing measures of **effect size.** The term "effect size" refers to the strength or magnitude of the statistical finding. Some effect sizes appear on a standardized scale, such as 0 to 1, so that researchers can compare them across different analyses, while others have "rules of thumb" that classify results into "small effects," "medium effects," and "large effects." For example, in the case of the t-test of two independent samples, statistician Jacob Cohen (1988, p. 67) developed the "d" statistic to provide a standardized measure of the size of the difference between two sample means. Cohen's d is conceptually simple: divide the mean difference by the pooled standard deviation of the two samples. As always, R provides multiple methods of obtaining statistics such as effect sizes. Here's one:

```
install.packages("effsize")
library(effsize)
cohen.d(mtcars$mpg[mtcars$am==0] ,mtcars$mpg[mtcars$am==1])
```

The output from the cohen.d() command shows an effect size of −1.47 and categorizes this effect as large. Cohen's d can be either positive or negative, depending upon the difference in the means, and is effectively calibrated in standard deviations. Our finding of $d = -1.47$ could be reported as "Cohen's d showed a large effect size of $d = -1.47$, indicating that manual transmissions were nearly one and a half standard deviations more fuel efficient than automatic transmissions." Try searching for "interpreting Cohen's d" and you will find a variety of rules of thumb for making sense out of this particular effect-size statistic. Generally speaking, researchers strive to design their studies to

The Calculation of t

As noted in the text, t is a family of distributions that can be used (among other things) as a model for the distribution of population mean differences that are estimated from sample data. Another way to think about this is that statisticians developed a theory that shows that when you repeatedly draw two samples and subtract the means, you get a sampling distribution that looks pretty normal except for having "thicker" tails. When the samples are small, say less than $n = 30$, the tails are quite thick. When the samples are larger, say over $n = 100$, then the t-distribution looks almost identical to the normal curve. The thicker tails come from the fact that we are somewhat uncertain of population variance when we estimate it from a sample.

Armed with that theoretical model, we can then use the means and standard deviations from two samples to calculate an observed value of t:

Independent-samples t:

$$t_{obs} = \frac{\left(\bar{x}_1 - \bar{x}_2\right)}{s_p \sqrt{\frac{1}{n_1} + \frac{1}{n_2}}}$$

This equation says that the observed value of t is equal to the difference in the sample means divided by the pooled standard error. The pooled standard error appears in the denominator of the right-hand side of the equation and is itself calculated by multiplying the pooled standard deviation by the two elements under the square root sign—these two elements combine the sample sizes of the two samples. I have not shown the calculation for the pooled standard deviation, but it adds together the variability of the two samples, making some adjustments for the combined sample size of the two samples.

When the t-test procedure has calculated the observed value of t from the sample data, as shown above, it then positions that value of t on the theoretical distribution of t values appropriate for the combined size of the two samples. For instance, with two samples, each of size $n = 10$, the appropriate t-distribution is on 18 **degrees of freedom**. The 18 degrees of freedom (df) comes from this simple calculation $df = (n_1 - 1) + (n_2 - 1)$. The position of the observed t-value on the t-distribution divides the distribution into regions (quantiles), just as we did in Chapter 3. The p-value represents all of the area in the tails of the distribution, beyond the observed t-value. In other words, p represents the probability of obtaining a value of t at least as high as what was actually observed.

The observant reader will realize that the $df = 18.332$ reported for the t-tests in Chapters 4 and 5 can't possibly be correct for samples of $n = 19$ cars with automatic transmissions and $n = 13$ cars with manual transmissions. We would have expected $(19–1) + (13–1) = 30$ degrees of freedom rather than the oddly fractional $df = 18.332$. The explanation is that the use of Welch's adjustment for unequal variances causes

(continued)

a new and smaller value of degrees of freedom to be used when the variances of the two samples do not match.

As we dig deeper into the null hypothesis significance test throughout the book, we will further discuss the meaning of degrees of freedom. For now, consider that in each of our two samples, when we calculated the sample mean, we created a situation where one of the observations in each sample could be mathematically inferred by knowing both the sample mean and all of the other observations. In other words, one degree of freedom was lost from each sample as a result of calculating the sample mean.

obtain the largest possible effect sizes—they want to maximize the differences between group means, as well as other kinds of statistical effects. The editors of many scientific publications encourage the publication of effect sizes alongside the results of null hypothesis tests to help put those test results in perspective.

A second concern about the NHST is that the foundational assumption, that is, the absence of any difference between the two groups, is fundamentally flawed. Working from our example of automotive transmissions, engineers worked on improving the efficiency of automatic transmissions for years, but in 1974 they knew full well that their designs were not as fuel efficient as manual transmissions. For researchers (or engineers) to start with a basic assumption that there would be no difference between the two types of transmissions makes no sense conceptually and in a practical sense gives us only a very low bar to cross in order to provide statistical support. Would it not make more sense to begin with the notion that there is some minimal useful difference between control and treatment groups and then use the data to assess the support for that minimal useful difference as the baseline? The Bayesian perspective, in particular, builds this essential assumption into the thought process from the beginning. From the Bayesian perspective, every new piece of data that we collect influences our thinking and refines our understanding in the context of our prior assumptions about the differences between the groups.

Researchers have addressed this NHST issue with a variety of approaches—too many to review here—but I will highlight one sensible idea from researcher John Kruschke (2013), who wrote the software for the BEST procedure we used above. Kruschke defines the "region of practical equivalence" (ROPE) as a range of values that the research can effectively consider equivalent to the null hypothesis. In our mtcars data, for instance, we might have stated, upfront, that if the automatic and manual transmissions were different from each other by anything less than 2 mpg, we would consider them effectively the same. So, while the NHST begins with the assumption of *exactly* 0 difference between the two means, we can choose to consider a wider interval around 0 as a more meaningful version of the null hypothesis—in this case anywhere between −2 mpg and +2 mpg. This strategy works comfortably with the

Bayesian approach—just see if the HDI overlaps with the ROPE. If the ROPE, representing a region of no meaningful difference, does not overlap with the HDI, then the probabilities argue in favor of the alternative hypothesis and the HDI provides the details regarding the range of possible differences.

The third concern about the NHST links the question of the choice of alpha level to the process of data collection and the investigator's choice as to when to begin statistical analysis. It is sensible at this point to give some more attention to what a *p*-value actually is. Formally speaking, a *p*-value is the probability—under the assumption of the null hypothesis—of observing a statistical result at least as extreme as what was actually observed. That's a mouthful, so let's break it down with reference to our mtcars example. The output of the t.test() command, as presented above, resulted in a *p*-value of .001374 based on an observed *t* of −3.7671. We can restate .001374 as 13.74/10,000 to make it easier to think about (in fact, let's just round that value to exactly 14/10,000). So in concrete terms, *if the null hypothesis were true*—exactly zero difference in reality between the two types of transmissions—*then only 14 out of 10,000 t-tests would yield a value of* t *larger in magnitude than 3.7671*. We can confirm this idea for ourselves using a random-number generator to create a distribution of *t*-values with a mean of 0 (similar to the sampling we did with rnorm() and runif() in Chapter 3):

```
set.seed(54321)                                # Control randomization
carsTdist <- rt(n=10000,df=18.332)             # 10,000 random t values
hist(carsTdist)                                # Show in a histogram
lowTvalues <- carsTdist[ carsTdist <= −3.7671] # Here is the lower tail
hiTvalues <- carsTdist[ carsTdist >= 3.7671]   # Here is the upper tail
# The number of observations in the tails
length(lowTvalues) + length(hiTvalues)
```

In this code, I use set.seed() to control the sequence of random numbers so that you can get the same result as I did. The set.seed() function controls the sequencing of random numbers in R to help with the development of code that contains random elements. If you use set.seed() at the beginning of a block of code, the same random numbers will be issued every time you run that code. Then I generate a random distribution of *t*-values using the degrees of freedom (18.332) shown by the *t*-test results above (also review "The Calculation of *t*"). When you run the histogram command you will find a nice symmetric distribution that looks normal except for slightly thicker tails. In the last three lines of code, I collected all of the low values of *t* (less than or equal to our observed value of −3.7671) and all of the high values of *t* (greater than or equal to 3.7671) and then I counted them. The result comes out exactly as we predicted: 14 extreme values of *t*, with half of them less than −3.7671 and half of them greater than 3.7671. By the way, you will note that we counted both extremely low values of *t* and extremely high values of *t*. This is because the recommended practice for the NHST is to always use a **two-tailed test.** By counting the probabilities in both tails, we are prepared to detect both unusually low values of t and unusually high values. Translated into our mtcars example, the t-test we conducted allowed us to detect situations where automatic transmissions were either significantly worse or significantly better than manual transmissions.

So we have confirmed that, under the assumption of the null hypothesis and with $df = 18.332$, the probability of observing a value of t with an absolute value greater than or equal to 3.7671 is .001374. This is well below the conventional alpha threshold of .05, so our NHST leads us to reject the null hypothesis of no difference between transmission types. But where did this alpha value of .05 come from and why is it the conventional threshold for evaluating the significance of a statistical test? It all goes back to the precomputer age. The alpha level sets the rate of so-called Type I errors—the probability of a false positive (the incorrect rejection of an accurate null hypothesis). Before there were computers, statisticians published long tables of critical values that researchers could use to look up t-values and other statistics to see if they were significant. Because these tables of **critical values** were so long, the authors had to compromise on how much detail they included. For example, in R. A. Fisher's (1925, p. 158) *Statistical Methods for Research Workers*, the table of critical t-values contains just three columns of critical values for t that Fisher considered significant: $p < .05$, $p < .02$, and $p < .01$. The use of $p <. 05$ (as well as $p < .01$, and occasionally, $p < .001$) became ingrained in the everyday practice of doing statistics (as well as teaching it) because every textbook and reference book included critical values for $p < .05$. It's kind of funny that today, when many people carry around mobile phones packed with computing power and high resolution displays, we stick with a strategy designed to save printing costs in 1920s textbooks.

REPLICATION AND THE NHST

Setting an alpha threshold of .05 means that we are willing to accept one false positive out of every 20 statistical significance tests that we conduct. Considering that many research articles conduct at least half a dozen different statistical tests, this suggests that at least one out of every three published articles contains at least one irreproducible result. Recent research on replications (repeating the procedures of a published piece of research and reporting the new result) and retractions (when a journal editor withdraws official support for a previously published article) suggests that these irreproducible results are quite common. Results published by Bohannon (2015) in *Science* suggested that perhaps fewer than 40% of articles published in the field of psychology could be replicated with similar statistical results and using the same research design. Anyone who has spent time conducting statistical tests on a small data set also knows that the addition or removal of just a few cases/observations can make the difference between $p = .051$ and $p = .049$, the former being nonsignificant and dooming one's research to the trash heap, versus the latter being significant and possibly eligible for publication. So, it is all too tempting to continue to collect data until such time as important null hypothesis tests reach thresholds of significance. There's a term for this that researchers have been talking about recently: *p*-hacking (Nuzzo, 2014). Researchers have at their disposal a wide range of methods for obtaining results that are statistically significant. Even in our simple mtcars example, there were many ways that we might have explored our research question about fuel efficiency and transmission type. We could have looked at domestic and foreign cars separately, only considered automobile models that had both a manual and automatic transmission option, tried to control for weight of the vehicle, and so forth. Before too long we might have at least 10 or more statistical tests, with a pretty strong chance of finding a false positive in our significance tests.

This is not to say that researchers are dishonest, but rather that our whole system of conducting and publishing scientifically studies has historically pushed researchers in the direction of seeking statistical significance rather than trying to understand at a deeper level the meaningfulness or practicality of what their data may say. Having accomplished all of the work of designing, setting up, and conducting a study, it is only human nature to want to publish the results. When editors and reviewers of scientific publications insist that *p* < .05 is the magic number, researchers create and conduct their studies in response to that goal.

There is also a problem at the other end of the telescope, that is, when researchers have access to large data sets instead of small ones. Once you have more than about *n* = 1,000 observations—a situation that is increasingly common in these days of big data—almost every difference, no matter how trivial, is statistically significant. In the context of big data, the NHST may not always provide useful information about the outcomes of a piece of research.

CONCLUSION

Building upon the idea of a confidence interval developed in the previous chapter, we ran our first Bayesian statistical procedure and reviewed a detailed distributional model of the mean differences between the two different types of transmissions in 1974 model year cars. The highest density interval (HDI) showed graphically where the mean difference in fuel efficiency was most likely to lie: a region surrounding –7.22 mpg and favoring the fuel efficiency of manual transmissions. Although these results seem similar to the idea of a confidence interval, there is an important difference: the Bayesian output showed a distributional model of the population parameter of interest, while the confidence interval provided just a single interval estimate of that parameter with no guarantee that the particular confidence interval we constructed actually contained the population parameter. The posterior distribution of mean differences also showed graphically that the idea of no difference between the two types of transmissions was not credible.

That implausibility is one of the flaws of the traditional technique that we next reviewed: the null hypothesis significance test (NHST). The NHST procedure uses a model that assumes zero difference between the two populations and seeks to find results so extreme as to make that assumption improbable. We reviewed several reasons why statisticians and researchers have begun to criticize the NHST, for instance, that rejecting the null hypothesis does not provide information about effect size. Even though the null hypothesis test is problematic, it is important to be able to make sense of research reports that use it. Even with all of the disadvantages, there are still many scientific journals that only report the results of null hypothesis tests.

When reading these articles, you can safely treat a declaration of statistical significance as a "minimum bar" that research results must cross on their way to relevance, meaningfulness, and/or practicality. You should always look for additional evidence of "practical significance" after you see that the results meet that minimal bar. In recently published work, you will be more and more likely to see confidence intervals mentioned. You may also see measures of effect size, which we will examine in additional detail as we look at specific tests that are applicable to a variety of research situations. As we look forward to future research, we may see more and more usage of Bayesian thinking and statistical procedures that provide definite detailed probability information concerning a range of possible research outcomes.

EXERCISES

1. In your own words, describe the process of Bayesian thinking. Make up an example that uses a two-by-two contingency table to document your ideas. Put real numbers in the contingency table and convert them to probabilities so that you can explain what happens when you focus attention on just one row or column.

2. Draw or plot a t distribution (hint: it looks like a normal distribution with thick tails). Don't worry about the degrees of freedom, but if you have to choose a value of df to create a plot, use $df = 18$. Mark the point on the t distribution corresponding to the null hypothesis and label it. Draw vertical lines to mark the start of the tails at the appropriate spots to leave 2.5% in the lower tail and 2.5% in the upper tail.

3. A scientist compares two groups of trees that have grown under different conditions. She measures and averages the heights of the trees within each group and conducts a t-test to compare them. There are $n = 10$ trees in both groups, so there are 18 degrees of freedom. The scientist calculates an observed value of $t = 2.2$, which has an associated p-value of $p = .041$. Comment on this result. What can you say about the implications of the study?

4. Using the same premise as Exercise 3, the scientist also obtains a 95% confidence interval, where the lower bound is 0.2 meters of difference between the two groups and the upper bound is 2.2 meters of difference. Comment on this result. What can you say about the implications of the study?

5. Using the same premise as Exercise 3, the scientist conducts a Bayesian comparison between the two groups. The procedure produces a highest density interval (HDI) plot. The mean difference at the center of the plot is 1.23 meters. The 95% HDI has a lower boundary of –0.1 meters and an upper boundary of 2.43 meters. The graph is marked to show that 2.2% of the population mean differences were lower than 0, while 97.8% of the mean differences were above 0. Comment on these results. What can you say about the implications of the study?

6. The PlantGrowth data set contains three different groups, with each representing various plant food diets (you may need to type data(PlantGrowth) to activate it). The group labeled "ctrl" is the control group, while "trt1" and "trt2" are different types of experimental treatment. As a reminder, this subsetting statement accesses the weight data for the control group:

PlantGrowth$weight[PlantGrowth$group=="ctrl"]

and this subsetting statement accesses the weight data for treatment group 1:

PlantGrowth$weight[PlantGrowth$group=="trt1"]

Run a *t*-test to compare the means of the control group ("ctrl") and treatment group 1 ("trt1") in the PlantGrowth data. Report the observed value of *t*, the degrees of freedom, and the *p*-value associated with the observed value. Assuming an alpha threshold of .05, decide whether you should reject the null hypothesis or fail to reject the null hypothesis. In addition, report the upper and lower bound of the confidence interval.

7. Install and library() the BEST package. Note that you may need to install a program called JAGS onto your computer before you try to install the BEST package inside of R. Use BESTmcmc() to compare the PlantGrowth control group ("ctrl") to treatment group 1 ("trt1"). Plot the result and document the boundary values that BESTmcmc() calculated for the HDI. Write a brief definition of the meaning of the HDI and interpret the results from this comparison.

8. Compare and contrast the results of Exercise 6 and Exercise 7. You have three types of evidence: the results of the null hypothesis test, the confidence interval, and the HDI from the BESTmcmc() procedure. Each one adds something, in turn, to the understanding of the difference between groups. Explain what information each test provides about the comparison of the control group ("ctrl") and the treatment group 1 ("trt1").

9. Using the same PlantGrowth data set, compare the "ctrl" group to the "trt2" group. Use all of the methods described earlier (*t*-test, confidence interval, and Bayesian method) and explain all of the results.

10. Consider this *t*-test, which compares two groups of $n = 100,000$ observations each:

   t.test(rnorm(100000,mean=17.1,sd=3.8),rnorm(100000,mean=17.2,sd=3.8))

   For each of the groups, the rnorm() command was used to generate a random normal distribution of observations similar to those for the automatic transmission group in the mtcars database (compare the programmed standard deviation for the random normal data to the actual mtcars data). The only difference between the two groups is that in the first rnorm() call, the mean is set to 17.1 mpg and in the second it is set to 17.2 mpg. I think you would agree that this is a negligible difference, if we are discussing fuel economy. Run this line of code and comment on the results of the *t*-test. What are the implications in terms of using the NHST on very large data sets?

CHAPTER 6

Comparing Groups and Analyzing Experiments

At this point in the book, we have reached an important point of departure. In previous chapters, we examined the basic logic of inference with the goal of understanding to what degree we can trust population parameter estimates obtained from a sample. To review, whenever we calculate a mean or any other statistic from a sample, we know that the result is not exactly the same as the underlying population that the sample is trying to represent. Depending on the size of the sample and other important concerns, we may or may not put a high degree of trust in the sample mean as an estimate of the population mean. Similarly, in a case where we have two sample means, and we compare them, we are estimating a difference in means between two underlying population groups. Statistical inference is a reasoning process that considers and quantifies the role of sampling error in estimates of means and other population parameters. From now on, we will apply the same inferential logic to a variety of other sample statistics.

In the previous chapter, we used a *t*-test and an equivalent Bayesian algorithm (BEST) to compare the means of two groups. Technically, the question we were asking pertained to the size of the mean difference between the two population groups that the samples represented. In this chapter we consider a more general case where we can simultaneously analyze the differences in means among any number of groups. This introduces a powerful new technique for comparing many different groups without having to repeat individual *t*-tests over and over again. Using this technique to analyze an experiment—for example, a comparison among the usability of different web pages—we do not have to boil everything down to pairwise comparisons. Using **analysis of variance** (ANOVA) we can examine combinations of different factors (e.g., comparing various colors on a web page, while also varying font sizes).

ANOVA belongs to a family of statistical techniques known as the general linear model, often abbreviated as GLM. Note that there is a very similar term, the genera*lized* linear model, that sometimes has the same abbreviation, but which has even more "generality" than the general linear model does. All of the members of the GLM family accomplish the same job: a set of **independent variables,** also often called **predictors,** are related to or used to predict a **dependent** variable, also called an **outcome variable.** Let's use an example to illustrate: we can model the cost of a tablet computer based on the operating system, the size of the screen, and the size of the memory. In this example, the cost of the tablet is the dependent variable and there are three independent variables: the categorical variable of operating system type (e.g., iOS vs. Android vs. Kindle), the size of the screen (which could either be a measurement or a designation of small/medium/large), and the memory size in gigabytes (GB). By using a GLM analysis on a data sample containing these four variables we can learn several interesting things:

- The overall quality of the statistical result, or you can also think of it as the strength or quality of prediction, represented by a single effect size value.
- An inferential statistical test on model quality, either to test the hypothesis that the statistical value (e.g., R-squared) is significantly different from zero (null hypothesis significance test), or to establish a likely range of values (confidence interval), or to show the likelihoods associated with different values (Bayesian test).
- One coefficient for each independent variable; these coefficients are used in a linear additive equation to model/predict the dependent variable.
- An inferential test on each of the coefficients, either to test the hypothesis that the values are significantly different from zero (NHST), or to establish a likely range of values (confidence interval), or to show the probabilities associated with different values (Bayes).
- Diagnostic information to make sure that all of the necessary assumptions about the data are satisfied; for example, that the error values from the predictions are normally distributed.

Sometimes statistics instructors teach ANOVA separately from other techniques in the GLM family (e.g., linear multiple regression). The only difference between ANOVA and these other techniques is that the independent variables in ANOVA tend to be **categorical** rather than **metric.** "Categorical" means that the variable indicates membership in two or more discrete categories. "Metric" means that the variable ascertains an ordered measurement of some type, such as ratings on a 10-point scale, weight, or number of bytes of memory. In our example of modeling tablet cost, the typical way of teaching ANOVA would be to begin by only modeling the operating system type and temporarily ignoring metric predictors such as memory size. Later in the book, when we get to the chapter on linear multiple regression, you will see that we can freely intermix various types of predictors, but for now we will go with the idea that ANOVA allows for comparing two or more separate groups of data, that is, an independent variable that is categorical.

ANOVA takes a somewhat different conceptual perspective on comparisons between groups than the *t*-test does. Instead of comparing means to accomplish the comparisons, we will be comparing variances. You probably remember from the beginning of the book that the variance is a measure of the "spread" of a variable and it is simply the square of the standard deviation. The core concept of ANOVA is that it uses two different variance estimates to assess whether group means differ. The technique partitions the overall variance among all of the observations into **between-groups** and **within-groups variance** to evaluate whether the samples might have come from the same underlying population. Don't worry if this does not make sense yet: we are going to dig in a lot deeper.

We can demonstrate this between-groups versus within-groups idea with a small data set. Let's begin by inventing a population and then sampling three different groups from it. We can use the built in "precip" data in R to create three groups. Note that the precip data contains precipitation amounts from 70 U.S. cities in a roughly normal distribution:

```
install.packages("datasets")
library(datasets)
set.seed(1)
pgrp1 <- sample(precip,20, replace=TRUE)
pgrp2 <- sample(precip,20, replace=TRUE)
pgrp3 <- sample(precip,20, replace=TRUE)
```

These commands create three randomly sampled groups of precipitation observations of 20 cities each. As I have done earlier in the book, I used the set.seed() function so that you will obtain the exact same results as me. Let's begin by looking at the variance for the data that we obtained in these three samples by concatenating the three samples together:

```
var(c(pgrp1,pgrp2,pgrp3))     # Join all of our sample data and calc variance
[1] 173.77
```

Now let's compare that value with the variance of the original population:

```
var(precip)
[1] 187.8723
```

As we would expect from any sampling process they do not match exactly. We have used a relatively small group size of *n* = 20 to get a total sample of *n*

= 60 and we know that the process of random sampling will compose a range of different samples even when all data are drawn from the same population. In the code above, we calculated the variance by using the raw data (number of inches of precipitation for each city in the sample) to calculate this statistic. But there is another way that we might look at variance: we could also look at the variability among the means of the three different groups, like this:

```
mean(pgrp1)     # Examine the means of the three groups
mean(pgrp2)
mean(pgrp3)
# Create a bar plot of the means
barplot(c(mean(pgrp1),mean(pgrp2),mean(pgrp3)))
var(c(mean(pgrp1),mean(pgrp2),mean(pgrp3)))  # Variance among the means
```

The result of the last command shows a variance of 1.78. That's quite a small number compared with either 173.77 or 187.8723, but not unexpected. If you feel unsure about why sample means drawn from the same population should have only a little variance among them, review the process of creating a sampling distribution as described in Chapter 3. Now let's play around with our data a bit and force one of the groups to be very different from the other two groups. Remember that the values in each of these groups represent the amount of annual rainfall in different U.S. cities. So let's pretend that our third group of cities experienced a terrible drought, thanks to some kind of weird climate change issue. We can make this happen by subtracting 5 inches of rain from every value in pgrp3:

```
# Take away five inches of rain from each point in sample 3
pgrp3 <- pgrp3 – 5
# Bar plot of the new means
barplot(c(mean(pgrp1),mean(pgrp2),mean(pgrp3)))
```

R knows how to do the right thing here by subtracting the *scalar* value of 5 from every member of the *vector* that is pgrp3. Let's examine the new bar plot of the group means. In Figure 6.1, the mean for the third group has diminished by 5 inches just as you would expect. There is now much more variation among the group means.

If we recalculate the variance among the means we get a radically larger result, nearly 10 times larger than the original variance among the means:

```
var(c(mean(pgrp1),mean(pgrp2),mean(pgrp3)))
[1] 17.53343
```

Yet if we rejoin the raw data from the three groups and recalculate the underlying population variance, we arrive at a value that is again very close to the original population value of 187.87:

```
var(c(pgrp1,pgrp2,pgrp3))
[1] 184.448
```

Does this result make sense to you? None of the individual rainfall values in our modified version of pgrp3 is especially unusual on its own, but it is remarkable to find so many low values gathered in one group. The group mean for pgrp3 now deviates from its fellow group means, even though the data points in that group are not unusual in the population as a whole. The artificial drought that we created in group 3 radically shifted the variance among the means, raising it by a factor of 10 from 1.78 up to 17.5, but when we mixed our modified raw data back in with the other two samples to create an overall estimate of the underlying population variance from the sample data, it had hardly shifted, rising from about 174 to 184, just a 6% increase.

This little simulation represents the basic logic of the analysis of variance. By pooling together all of the data we have collected from all of the different groups, we can create a reasonable estimate of the underlying population variance, under the assumption that all of the data from all of the groups was drawn from the same population. Then we can use the variance among the means as another way of looking at the data. If the groups have all been sampled from the same underlying population, then the *variance among the means* of those groups will be proportional to the estimate of the population variance. If one or more of the groups have been sampled from a population with a very different mean, however, the variance among the sample means will tend to exceed what we would expect if the samples all came from the same population.

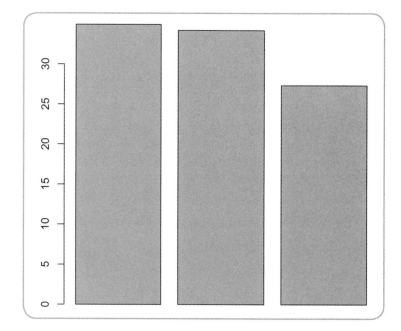

FIGURE 6.1. Three group means after modifying the third group.

Formulas for ANOVA

Before statistical software was widely available, statistics instructors routinely taught students both definitional and calculation formulas for common statistical quantities. The calculation formulas helped to speed the process of calculating a statistic from raw data when all you had was a pencil and paper. As we have R to help us with all of our statistics, those calculation formulas are not especially useful. The definitional formulas, however, still have some value in that they help to illuminate some of the concepts that underlie the statistics. In the case of ANOVA, there are three such formulas that show quite neatly how the variability within a data set can be reckoned as a between-groups portion (the variability among means) and a within-groups portion (the variability of scores with respect to their group means). Each of the following formulas represents variability in terms of the sum-of-squared deviations from a mean:

Total Sum-of-Squares: $$SS_{\text{total}} = \Sigma\left(x - \overline{G}\right)^2$$

This formula says that total sum-of-squares in the data set is the sum, across all scores (x), of the squared difference between each score and the **grand mean**. The grand mean is the mean of all of the scores in the whole data set, that is, ignoring the grouping variable. This formula might seem very familiar because it is at the heart of calculating the regular old variance and the standard deviation.

Next, let's look at the within-groups variability:

Within-Groups Sum-of-Squares: $$SS_{\text{within}} = \Sigma\Sigma\left(x_{ij} - \overline{X}_j\right)^2$$

This formula states that the within-groups sum-of-squares consists of the sum of the squared deviations of each score from its respective group mean, with the results from all the groups summed together. I've left out a mess of subscripts to keep things simple, but the right hand summation symbol (Greek capital letter sigma) does the summing within each group and the left hand summation symbol adds together the results from each group. For example, if this data set had three groups, we would divide up the scores into those three groups and calculate a mean for each group. Then we would use each group mean to calculate a separate sum of squared deviations for the scores within each group. Then we would add together those three separate sums.

Finally, here is the between-groups variability:

Between Groups Sum-of-Squares: $$SS_{\text{between}} = \Sigma n\left(\overline{X}_j - \overline{G}\right)^2$$

This formula calculates the squared deviation between each group mean and the grand mean and adds the results together. For example, if we had three groups

(continued)

we would be adding three squared deviations together. Note that after calculating each squared deviation, the result is multiplied by the number of observations in that particular group. That scaling factor puts the between-groups variability on the same footing as the within groups variability. In fact, if you add together the sum-of-squares within groups to the sum-of-squares between groups, you will get the total sum-of-squares again, like this:

Total Sum-of-Squares: $SS_{total} = SS_{within} + SS_{between}$

In other words, we have just partitioned the total variability in the data set into two components, the variability represented inside of each group (summed across groups) and the variability among the group means (with the appropriate scaling factor). In this chapter, when we calculate a new statistic known as F, the variances that go into that calculation come right from the sum-of-squares values explained above.

In fact, we can create a new statistic, a ratio of variances. The numerator of the ratio is the scaled variance among the means of the groups—between-groups variance. The denominator of the ratio is the variance within the groups (pooled together to form a single estimate)—naturally called the within-groups variance. This ratio is called the "F-ratio" and any F-ratio that is substantially larger than 1.0 is considered possible evidence that at least one of the groups is from a population with a different mean. That's because both the numerator and the denominator represent different methods of estimating the same thing—the population variance. If the between-groups variance is considerably larger than expected, it suggests that the differences among the group means are bigger than what would be expected.

As you know, every time we sample from a population, we get slightly different data. That means we will also get different values for within-groups variance. If we divide our sampled data into groups we will also get various group means and slightly different variances among those group means each time we sample. Put this altogether in your head and you may realize that we will get a slightly different F-ratio every time we sample. Under the assumption (the null hypothesis) that we are sampling all of the groups from the same population, though, most of our F-ratios should be very close to 1.0. We can double-check that idea by asking R to generate a random list of F-ratios. These commands generate a list of 100 F-ratios. I have set the parameters to match the original precipitation scenario, where each of the groups (and therefore each group mean) was sampled from the population (without the later tinkering that we did with group 3). Figure 6.2 displays a histogram of the F-values:

```
randomFs <- rf(n=100,df1=2,df2=57)
hist(randomFs)
```

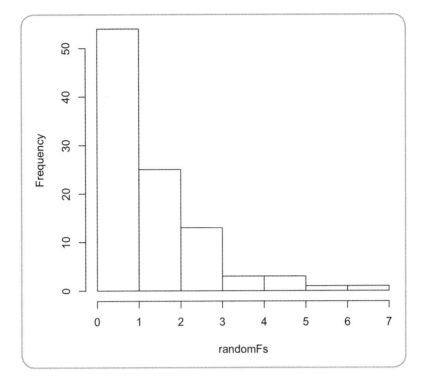

FIGURE 6.2. A set of 100 randomly generated *F*-ratios.

Take a close look at Figure 6.2. The great majority of these *F*-ratios are right near 1, just as we predicted. Put into words, if we draw multiple groups from the same population, the scaled between-groups variance and the within-groups variance will generally be about equal to each other. Notice though, that in about five of these cases (look above the *X*-axis values between 4 and 7 in Figure 6.2), we have an unexpectedly large *F*-ratio. In these situations, *just because of sampling error*, the means of the groups have diverged from one another, making the scaled between-groups variance much larger than the within-groups variance. These are false positives caused by sampling error. One last thing: the command that we used to generate this bunch of *F*-ratios, rf(), had a couple of arguments, df1 and df2, that we will discuss in some detail below. I chose df1 = 2 and df2 = 57 to match our precipitation example, as you will see.

FREQUENTIST APPROACH TO ANOVA

So let's move on to trying out these ideas about variance using some statistical procedures in R. In this case, we will begin with the traditional frequentist methodology—in other words, the Null Hypothesis Significance Test (NHST). As with most techniques, R provides multiple ways of accomplishing the same thing. For our first ANOVA we will use a function called aov() which is a wrapper around another function called lm(), which in turn is an implementation of the general linear model using a very efficient computational technique known as least-squares fitting. I will discuss least squares in more detail in later chapters, but the main thing to know for now is that it is a mathematical calculation that finds ideal coefficients or "weights" that connect the independent variables to the dependent variable. Here's all the code we need to sample a brand new set of 60 observations from the "precip" data set, set up three groups, check their distributions, run an ANOVA, and check the output:

```
# Run ANOVA on groups sampled from the same population
set.seed(10)              # Control the randomization
# Enough for 3 groups of 20
precipAmount <- sample(precip,60,replace=TRUE)
# Group designators, 3 groups
precipGrp <- as.factor(rep(seq(from=1,to=3,by=1),20))
# Put everything in data frame
precipDF <- data.frame(precipAmount, precipGrp)
# Get a box plot of the distribs
boxplot(precipAmount ~ precipGrp, data=precipDF)
# Run the ANOVA
precipOut <- aov(precipAmount ~ precipGrp, data=precipDF)
summary(precipOut)     # Provide an ANOVA table
```

I have added comments to this code block to help you navigate. The set.seed() function, as noted previously, controls the sequence of random numbers to help with code debugging and to ensure that you get the same sampling results as me. Next, we use sample() to draw 60 data points at random from the built-in precip data set. For this first exploration of ANOVA, *we want to show what happens when all of the data are sampled from the same population.* The rep() and seq() commands generate a repeating sequence of 1, 2, 3, which we will use as group designators. If you want this made-up example to be more concrete, you could think of group 1 as the precipitation in East Coast cities, 2 as the Midwest, and 3 as West Coast. Then we use data.frame() to bind the two variables—the precipitation data and the group designators—into a data frame. We check our result with a boxplot (see Figure 6.3) that shows the medians of the three distributions as very similar, even though the distributions differ from each other to some degree.

The aov() command runs the ANOVA. Note that the first argument to aov() is this string that describes the model we want to test: "precipAmount ~ precipGrp." Expressed in words, this notation simply says: Test precipAmount as the dependent variable and make it a function of the independent variable(s) that follow the "~" character (in this case just the grouping variable precipGrp). The aov() command produces output that we place in a variable called precipOut. Finally, we run the summary command on this output object. The summary() provides a standard ANOVA table, which looks like this:

	Df	Sum Sq	Mean Sq	F value	Pr(>F)
precipGrp	2	90	45.11	0.247	0.782
Residuals	57	10404	182.53		

This looks a little confusing at first, but it is a basic ANOVA table, and learning to interpret it is an important skill. The output abbreviates key terminology for the analysis as follows:

• df—degrees of freedom: A statistical quantity indicating how many elements of a set are free to vary once an initial set of statistics has been calculated; from a data set of 60 we lose one degree of freedom for calculating the grand mean; among the three group means only two can vary freely; this leaves 57 degrees of freedom within groups (aka residuals);

• Sum Sq—sum of squares: A raw initial calculation of variability; the first line is the "between-groups" sum of squares discussed above; the second line is the "within-groups" sum of squares (see Formulas for ANOVA on page 93);

• Mean Sq—mean squares, the sum of squares divided by the degrees of freedom, aka variance: the first line is the "between-groups" variance as discussed above; the second line is the "within-groups" variance;

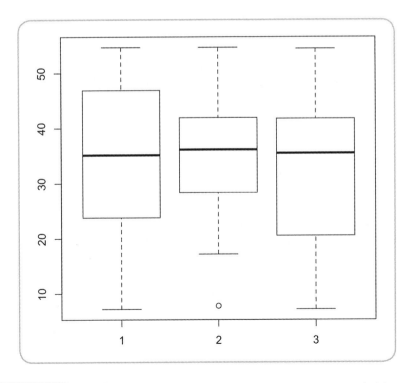

FIGURE 6.3. Box plots for the three precipitation groups sampled from one population.

• F-value—the *F*-ratio as discussed above: quite literally a ratio of the mean squares from the first line (between groups) and the mean squares of the second line (within groups), that is, 43.47 divided by 180.19;

• Pr(>F)—the probability of a larger *F*-ratio: when examining a random distribution of F-ratios for the degrees of freedom appearing in this table, this is the probability of finding an *F*-value at least this high (in this case at least 0.247, which is a really small *F*-value). The *F* distribution only has a positive tail, so for us to reject the null hypothesis, we must look for extreme values of *F* that appear in the tail of the distribution.

• precipGrp—the grouping variable for the ANOVA, also known as the independent variable, just as we specified in the call to aov(). The line of results to the right of this label represents pretty much everything we are trying to find out from this ANOVA.

• Residuals—this line accounts for all of the within-groups variability. A residual is what is left over when all of the systematic variance is removed—in this case everything that is left over after "precipGrp" is taken into account. In some textbooks and other statistics programs, this line is sometimes labeled "Within Groups."

So, what to make of these particular numbers? Everything to the left of the *F*-value is simply presented as a way of cross-checking the results. The two most important values are the *F*-value itself, and the probability of finding a larger *F*. If you recall the discussion at the beginning of this chapter, you will remember that when the data from all of the groups is sampled from the same underlying population, we expect the between-groups variance (mean square) to be roughly equal to the within-groups variance (mean square). The former is calculated based on the spread of the means, while the latter is calculated based on the variance in the raw data, when all of the raw data from the three groups is pooled together. For an ANOVA result to be statistically significant, *F* must substantially exceed one. The Pr(>F) is the significance level, in other words the probability of finding a value of *F* larger than the observed value under the assumption that all of the data were sampled from the same population (the null hypothesis). For *F* to be significant according to the logic of the null hypothesis significance test, the value of Pr(>F) must be less than the alpha level chosen by the experimenter (typically, 0.05, 0.01, or 0.001) before conducting the test. In this particular case, $p = 0.782$, is much larger than any of these conventional alpha values, so in frequentist terms we have *failed to reject the null hypothesis*, where the null hypothesis was that all three groups were sampled from the same population.

Thus, the results of this test turned out just as we expected. We sampled all three groups from the same underlying population, so the value of *F* should be small and its *p*-value large. Sampling all three groups from the exact same

More Information about Degrees of Freedom

Our old friend William Sealy Gosset documented a problem in his 1908 paper, "The Probable Error of a Mean." In the introduction to the paper (p. 1), he mentioned, "as we decrease the number of experiments, the value of the standard deviation found from the sample of experiments becomes itself subject to an increasing error." This insight led to the application of $(n − 1)$ instead of n in the denominator of the calculation of the variance, when estimating the variance of a population from the variance of a sample:

Sample variance:
$$s^2 = \frac{\Sigma \left(x_i - \bar{x}\right)^2}{\left(n - 1\right)}$$

This formula says that the sample variance is a fraction where the numerator contains the sum of all squared deviations from the sample mean and the denominator is 1 less than the sample size. The use of $(n–1)$ naturally makes the variance slightly larger than it would be if you used n in the denominator, suggesting that if we were to (incorrectly) use n instead, it might lead to an underestimation of population variance. What is the cause of this underestimation? Try this little chunk of code to find out:

```
install.packages('gtools')      # install gtools to get permutations
library(gtools)                 # Make the package ready
tinyPop <- c(1,2,3)             # Here is a tiny population
# This next command gives every possible sample with replacement!
allSamp <- permutations(n=3,r=3,v=tinyPop,repeats.allowed=T)
allSamp                         # Verify: 27 unique samples
apply(allSamp,1,var)            # List sample variance of each sample
mean(apply(allSamp,1,var))      # What is the mean of those variances?
```

In this code, we use a package called gtools to get a permutation command that allows us to list all 27 of the possible samples (with replacement) of a tiny population that consists of just three elements, namely, the numbers 1, 2, and 3. Then we calculate the sample variance of each of those samples: the var() command (used here inside of a call to apply() that repeats the calculation for all 27 samples) properly uses the $(n–1)$ denominator. Then we average those variances: The result is 0.6666667.

Now the actual, true, correct *population* variance for our three–element population, tinyPop, is easy to calculate, even in your head: the mean is 2, so the sum of squares is $(1–2)^2$ plus $(2–2)^2$ plus $(3–2)^2$ also known as 2. Divide the sum-of-squares by the size of the population: $2/3 = 0.6666667$ to get the variance. So, calculating the population variance (correctly) with n in the denominator leads to the same result as using $(n–1)$ in the denominators of the sample calculations, when we

(continued)

average across all samples. This is what is meant when statisticians call the sample variance (with denominator $n-1$) an **unbiased estimator.**

The reason we need $n-1$ instead of n is that for each sample, we calculated its variance with respect to the *sample* mean, rather than the population mean, biasing the result. In 20 out of our 27 samples, the sample mean was wrong (i.e., not equal to the population mean)—either too low or too high—making the variance incorrect as well. When we calculate that sample mean in order to use it for other things (such as estimating population variance), we are therefore capitalizing on the uncertainty intrinsic in the sample. Using $(n-1)$ in the denominator of the variance calculation corrects for this uncertainty.

When we calculate the sample mean, we are effectively borrowing information from the sample. We can't manufacture new information from the sample, so the mean doesn't represent anything new. Consider our tinyPop variable above: If I told you that the mean was 2 (which it is) and that two of the data points were 1 and 3 (which they are), you could easily tell me that there is no choice whatsoever about the third data point. It absolutely has to be 2 if the mean is also going to be 2. So, the mean borrows one degree of freedom from the sample—one element in the sample can no longer vary freely once the sample mean is known. All of the downstream calculations using the sample mean have to somehow reflect that borrowing. In most of the rest of the book, if you see an expression such as $df = n-1$, what you are seeing is that we have used up a degree of freedom from our sample in order to obtain an unbiased estimate of some other statistic.

population is by and large the very definition of the null hypothesis, so the F-test did its job perfectly in this case. Remember that if we repeated this process 1,000 times, just occasionally we *would* get results that appeared statistically significant just by chance. Sampling errors cause some results to look significant even when they are not (a false positive; also known as a Type 1 error). As a reminder from Chapter 5, if we set our alpha level to 0.05, this suggests that when the null hypothesis is true, we will nonetheless observe a significant F-value in 5% of our samples.

Before we move on to an additional example, let's consider why the F-test has two different degrees of freedom attached to it. In the precipitation data set we are working with here, we started with 60 observations and lost one degree of freedom when we calculated the overall mean of the data set (total $df = 59$). Out of that total $df = 59$, we borrow two degrees of freedom to represent the between-groups variance. Why two degrees of freedom when we have three groups? Consider that if we know the grand mean and the means of two of the three groups, the third mean is no longer free to vary: we could calculate it directly from the other information we have. So generally speaking, if we have k groups, the degrees of freedom between groups is $k-1$.

With total $df = 59$ and between-groups borrowing two df from the total, the remaining degrees of freedom are allocated to the within-groups variance ($df = 57$). Together, between-groups df and within-groups df always add up to total df (in this case $2 + 57 = 59$). Like the t-distribution, the F-distribution is a family of distributions each of which has a slightly different shape. In contrast to the t-distribution, the exact shape of F depends upon both the between-groups and the within-groups degrees of freedom. When we called aov() it figured out all of the degrees of freedom for us and presented it in the ANOVA table. When aov() calculated the p-value for the null hypothesis test, it looked up the observed F-value in the F distribution on 2 and 57 degrees of freedom—between- and within-groups df, respectively. You may wish to explore the shapes of different F distributions by starting with the following code:

```
fVals <- seq(from=0.01,to=5,by=0.01)     # Make a sequential list of F values
# Plot the density function for this scenario
plot(fVals, df(fVals,df1=2,df2=57))
# Add points to the same plot, different df1
points(fVals, df(fVals,df1=3,df2=57))
# Add points to the same plot, different df1
points(fVals, df(fVals,df1=4,df2=57))
```

In the first command, we generate a list of F-values ranging from near 0 to 5. Then, using the density function for F, called df(), we plot the likelihood for each value of F using between-groups df=2 and within-groups df=57. Make sure to run the last two commands one at a time: in those commands we add points that describe curves for different values of df1 (between groups df) so you can compare the shapes of the curves. You can change the df1 and df2 parameters to any values you want in order to explore the effect of having more groups or more raw data. The density function shows us something similar to what we would get from creating a histogram of zillions of random F-values: each point along the density curve represents the proportion of F-values one would observe at a given level of F, under the assumption of the null hypothesis. To think about the p-values used in a null hypothesis test, mark a vertical line somewhere on the X-axis and consider how much area there is under the right-hand tail. I marked my vertical line at about $F = 3$ and it looked like the tail contained about 5% of the area: That's the $p < .05$ that we're talking about when we set the alpha level for a null hypothesis test.

Recall that when we did the t-test in Chapter 5, we had a measure of effect size called Cohen's d. For ANOVA, some statisticians prefer a quantity called *eta-squared*. Eta-squared can easily be calculated from the ANOVA table, which I insert again here as a reminder:

	Df	Sum Sq	Mean Sq	F value	Pr(>F)
precipGrp	2	90	45.11	0.247	0.782
Residuals	57	10404	182.53		

Eta-squared is calculated as the between-groups sum of squares divided by the total sum of squares. In this case eta-squared would be 90/(90+10404). If you compute this you will find that it comes out to 0.0086, a very small value. This value can be interpreted as the proportion of variance in the dependent variable (precipAmount) that is accounted for by the independent variable (precipGrp)—so in this case it is a little less than 1%. The real usefulness of eta-squared comes into play when the ANOVA table has multiple effects represented in it, and we can then calculate eta-squared values for each of the individual effects.

THE BAYESIAN APPROACH TO ANOVA

In the frequentist approach described above, we use the data presented to us as a way of amassing evidence to try to reject the null hypothesis. Specifically, we used the data to estimate the variance of a population, under the assumption that all of the subsamples—the group data—were drawn from a single population. Then we use the means of the groups to reestimate the population variance. Finally, we created a ratio of those two estimates, a test statistic called the F-ratio. We used a theoretical distribution of the F-ratio to place the observed test statistic in either a "highly unusual" zone (with an observed p-value smaller than 0.05, 0.01, 0.005, or 0.001 depending on a preselected alpha threshold) or a "pretty typical" zone (with an observed p-value larger than a preselected alpha). Any value of the F-ratio that falls in the "highly unusual" zone is taken as evidence to reject the null hypothesis. Alternatively, if we obtain any value of the F-ratio in the "pretty typical" zone, we fail to reject the null hypothesis.

As you look back over the previous paragraph, you will see an elaborate chain of reasoning that really tells us nothing at all about the probability that the null hypothesis adequately represents the data or, alternatively, the probability that an alternative hypothesis does a better job. Using Bayesian thinking, we can use our data to analyze these probabilities, and thus have more straightforward metrics of the strength of our belief in the null hypothesis and/or the alternative hypothesis.

We must begin by installing and loading the new R package we will use for this analysis, known as "BayesFactor." Like the BEST package we used in Chapter 5, BayesFactor is a new, innovative package. When I initially tried to install it, I found that the main software repository I used did not have a copy of the code for this package. I had to update my version of R to the latest and choose a different software repository. Hopefully you will not encounter this problem, but be alert to any error messages that come from running these two commands:

```
install.packages("BayesFactor")
library(BayesFactor)
```

Giving Some Thought to Priors

If you have your Bayesian cognitive crown securely affixed to your cranium, you may have been wondering about our *priors*. Remember from "Notation, Formula, and Notes on Bayes's Theorem" elsewhere in the book that Bayesian thinking starts with some prior probabilities about the situation that we modify using data to develop some posterior probabilities. The essential logic of Bayesian thinking is that the prior probabilities are the starting point, the data we have collected shifts our beliefs about the prior probabilities, and the result is the posterior probabilities. Yet, in the Bayesian *t*-tests we conducted in Chapter 5, as well as the Bayesian ANOVA in this chapter, we don't seem to have actually set any priors.

Statisticians have been discussing questions of whether and how to ascertain prior probabilities when little is known about the question one is trying to address with statistical analysis. One group has argued that as long as you have a sizable data set, the prior probabilities are not especially influential on the results. For example, if your friend who is about to toss a coin a few times just got back from the joke shop, you might have a prior belief in a fair coin of only 10%. That's a strong prior, and if the first few tosses all came up heads, reinforcing your belief, the posterior probability for a fair coin would also be low. As your friend continues, however, you now begin to see about half heads and half tails emerging. After a few dozen tosses showing roughly half heads and half tails, the posterior probability of a fair coin starts to rise toward 1. As we collect more and more data suggesting a fair coin, the weight of evidence starts to overwhelm the prior beliefs.

This insight has led some statisticians toward the idea that **uninformative priors** may suffice except when working with small samples of data (Efron, 2013). If you remember the discussion of Markov chain Monte Carlo, you will remember that we sample repeatedly to establish a posterior distribution of the population parameter(s) of interest. Rather than stating a prior belief explicitly (as we did above about the fair coin), we can also sample from a distribution of priors that represents a range of beliefs. In fact, if we chose to sample a value from the uniform distribution, we would essentially be saying that we have no clue what the prior probability should be.

In some cases, however, we may not be quite so clueless. In the case of the BEST procedure that we examined in Chapter 5, the priors for the means of the two respective populations are sampled from normal distributions. Kruschke (2013) termed these as **noncommittal priors**, signifying that they are perhaps more meaningful than uninformative priors, but without making a commitment to specific values. Although the jury is still out on whether uninformative, weakly informative, or noncommittal priors can consistently lead to good statistical conclusions, throughout this book we will typically use the default settings offered by package authors.

In the Bayesian ANOVA procedures used in this chapter (from the R package known as "BayesFactor") we use priors from the so-called Cauchy distribution to

(continued)

seed the analysis process (Morey, Rouder, & Jamil, 2013). The Cauchy distribution models the ratio of two variables—if you think about it, that is perfect for ANOVA, where we are trying to make sense of the ratio of between-groups variance to within-groups variance.

If you are curious about the effects of priors, you could easily do some experiments using the data and procedures introduced in Chapter 5. For example, the following code compares automatic and manual transmissions using the BESTmcmc() *t*-test, with the prior mean and standard deviation for both transmission groups set to plausible and identical values (mean = 20, *SD* = 4):

```
install.packages("BEST")      # May not need this if BEST is downloaded
library(BEST)                 # Likewise, may not need this
data(mtcars)                  # Makes mtcars ready for analysis
priorList <- list(muM = c(20,20), muSD = c(4,4))     # 1 mean, 1 SD for each group
carsBest2 <- BESTmcmc(mtcars$mpg[mtcars$am==0],
    mtcars$mpg[mtcars$am==1], priors=priorList)
plot(carsBest2)               # Review the HDI and other results
```

Try out this code to find out how much the posterior values have changed from the results you review in Chapter 5. Would you change your conclusions about the research question? Also try experimenting with prior values different from those provided above. What if you set the prior mean for the automatic transmissions *higher* than the manual transmissions? How much would you have to change the priors in order to draw a different conclusion from the data?

The BayesFactor package also has a number of "dependencies," that is, other packages it relies upon, such as the "coda" package. In most cases these will also download and install for you automatically, but you should be alert to any warnings or error messages that may appear in the console. The anovaBF function we will run uses the same formula notation to describe the model as the aov() function did above. Here is the command:

```
precipBayesOut <- anovaBF(precipAmount ~ precipGrp, data=precipDF)
mcmcOut <- posterior(precipBayesOut,iterations=10000) # Run MCMC
    iterations
plot(mcmcOut[,"mu"]) # Show the range of values for the grand mean
```

As with the aov() command earlier in this chapter, we have placed the output of the analysis into a data object called precipBayesOut. You can run str() on this object to look at its internal structure. In the subsequent command, we generate posterior distributions for the population parameters estimated by this procedure and place those in an object called mcmcOut. If you hearken back to Chapter 5, you will remember that for two groups we estimated the parameters for the mean of each group and then subtracted the mean of the second group from the mean of the first group to get the mean difference. In this case, we obtain an estimate for the grand mean in the population and then separate estimates of deviations from the grand mean for each of the three groups. The final command above provides a trace plot and a density plot (histogram) for the grand mean. The result appears in Figure 6.4.

The trace plot and the density histogram shown in Figure 6.4 actually display the same information in two different forms. The density histogram on the right side of the display follows the same logic as the graphical output of the BEST procedure from Chapter 5: it shows a histogram of the range of population values obtained from a Markov chain Monte Carlo (MCMC) analysis of the posterior distribution of the grand mean. The left hand side of the display shows a trace plot that captures each of the 10,000 steps of the MCMC run. This is helpful as a diagnostic: As long as there are no extreme outliers in this display (which there aren't) and the variation is consistent across the whole run (which it is), we can feel confident that this MCMC run converged on a stable result.

Of course, when considering the results of an ANOVA analysis, the grand mean is simply a reference point and what we are most interested in is how far

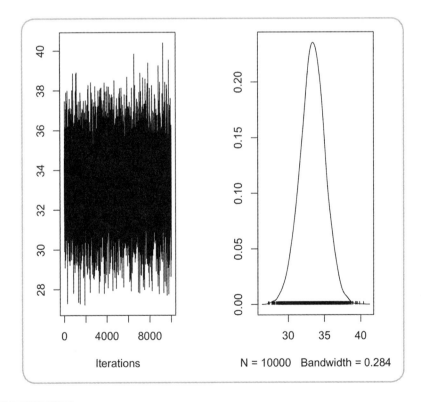

FIGURE 6.4. Trace plot and density histogram for the posterior population distribution of the grand mean.

each of the group means deviate from the grand mean (and, by extension, from each other). We can create a display that shows the 95% highest density interval for each group, showing the posterior distribution of its deviations from the grand mean. The following commands create that display for the first group, which appears in Figure 6.5.

```
par(mfcol=c(1,1))              # Reset display to one column if needed
hist(mcmcOut[,"precipGrp-1"])       # Histogram of MCMC results
# Mark the upper and lower boundaries of the 95% HDI with vertical lines
abline(v=quantile(mcmcOut[,"precipGrp-1"],c(0.025)), col="black")
abline(v=quantile(mcmcOut[,"precipGrp-1"],c(0.975)), col="black")
# Give numeric value of lower bound
quantile(mcmcOut[,"precipGrp-1"],c(0.025))
# Give numeric value of upper bound
quantile(mcmcOut[,"precipGrp-1"],c(0.975))
```

The histogram shown in Figure 6.5 has its center near 0. The lower and upper bounds of the 95% highest density interval (HDI) are marked with vertical lines −3.82 and 4.52, respectively. You can interpret this outcome as suggesting that the mean of precipitation group 1 does not deviate meaningfully from the grand mean: not only does its HDI straddle 0, but the region near 0 has

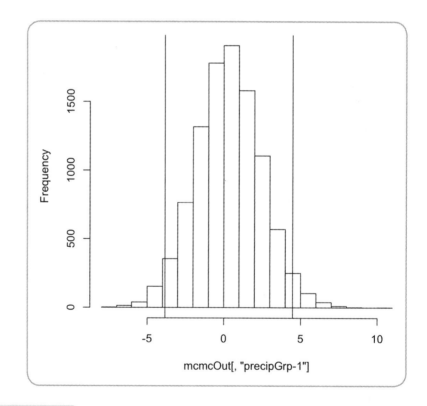

FIGURE 6.5. Highest density interval for the first precipitation group.

the highest probability of containing the population parameter. As an exercise you could develop your own HDI graphs for participation groups 2 and 3, but it might be more meaningful to compare the three distributions on the same graph. Luckily we have the boxplot() command that can accomplish that for us. Figure 6.6 displays the results:

```
boxplot(as.matrix(mcmcOut[,2:4]))
```

The resulting boxplot, as shown in Figure 6.6, summarizes the posterior distributions of all three groups together. Remember that these distributions represent deviations from the grand mean, which we discussed earlier and plotted in Figure 6.4. Also remember that the black bar in the middle of each box represents the median and the box itself contains the central 50% of all observations. Not only do all three boxes straddle the 0 point (i.e., no difference from the grand mean), but the whiskered areas for each group also overlap each other, suggesting no differences among the group means. Once you have made sense of all of the graphics, it will become easier to review the numeric results of the MCMC sampling, which you can accomplish with this command:

```
summary(mcmcOut)
```

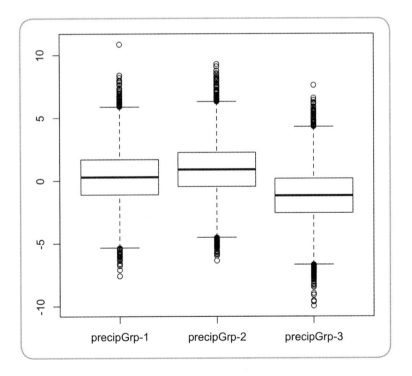

FIGURE 6.6. Box plot comparing the posterior distributions of deviations from the grand mean for three groups sampled from the same population.

This command provides the following results on the R console:

Iterations = 1:10000, Thinning interval = 1
Number of chains = 1, Sample size per chain = 10000
1. Empirical mean and standard deviation for each variable,
   plus standard error of the mean:

	Mean	SD	Naive SE	Time-series SE
mu	33.4184	1.756	0.01756	0.01756
precipGrp-1	0.2952	2.117	0.02117	0.02117
precipGrp-2	0.8990	2.123	0.02123	0.02123
precipGrp-3	-1.1942	2.094	0.02094	0.02127
sig2	183.8660	35.405	0.35405	0.37293

2. Quantiles for each variable:

	2.5%	25%	50%	75%	97.5%
mu	29.93080	32.23848	33.4093	34.6090	36.834
precipGrp-1	-3.82147	-1.09455	0.2705	1.6718	4.520
precipGrp-2	-3.29773	-0.49208	0.8943	2.2766	5.111
precipGrp-3	-5.42499	-2.53625	-1.1764	0.2009	2.903
sig2	127.65038	158.84900	179.1114	203.7983	265.643

Based on your work in Chapters 4 and 5, you are ready to make sense out of this output. The output is conveniently organized into two sections: Section 1 is entitled "Empirical mean and standard deviation for each variable, plus standard error of the mean." Each of these means is the mean value of the posterior distribution of the respective population parameter across the 10,000 samples that posterior() drew using MCMC. As mentioned above, our model focuses on estimating the grand mean and each group's deviation from the grand mean. We also get the population variance (denoted sig2, which stands for sigma-squared), which it is necessary to model in order to put the variations among the group means into context. The column marked SD shows the standard deviation for each parameter, also calculated across the 10,000 sampled posterior values. The Naive SE is the "naïve" standard error, which is obtained simply by dividing the SD by the square root of 10,000. The Time Series SE is calculated in a slightly different manner from the MCMC run.

While those statistics are handy for orienting ourselves to the results, the information in Section 2 will help us draw inferences about the differences among the groups. Section 2 is entitled "Quantiles for each variable," and these quantiles conveniently contain the lower and upper bounds of the 95% Highest Density Interval for each parameter. The left-most column, entitled "2.5%," represents the lower bound of the HDI and the right-most column, entitled "97.5%," shows the upper bound of the HDI. For any pair of the groups to be credibly different from one another, their respective HDIs must not overlap. When you inspect the output for Section 2, you will see that each of the three groups overlaps quite substantially with the other two, so there are no credible differences among these groups. Don't forget that these values represent distributions of deviations from the grand mean. To that end, you can also confirm from the output what we saw in Figure 6.6, that the 95% HDIs all straddle 0. This signifies that none of the group means is credibly different from the grand mean and supports our inference that these three groups were all drawn from the same population.

All of that rich information we obtained from examining the posterior distributions of the population parameters is what we have come to expect from the Bayesian approach to inference, but the BayesFactor package has one additional trick up its sleeve. It is called the BayesFactor package because the authors (Rouder, Speckman, Sun, Morey, & Iverson, 2009; Rouder, Morey, Speckman, & Province, 2012) built their approach on top of work originated by mathematician Harold Jeffreys (1891–1989) and later updated by statisticians Robert Kass and Adrian Raftery (1995). Jeffreys' classic book, *The Theory of Probability* (1998, first published in 1939) introduced the idea of a **Bayes factor.** The concept of a Bayes factor is straightforward: it is nothing more or less than an odds ratio that results from the comparison of two statistical models. For the inferences we have been considering, one of the simplest model comparisons we might accomplish lies in comparing the likelihood of a particular research hypothesis with a null hypothesis. Thinking back to Chapter 5, our research hypothesis was that manual transmissions and automatic transmissions differed with respect to fuel efficiency. Our null hypothesis was that there was a negligible difference between the transmission types. What if the odds of the research hypothesis (in comparison to the null hypothesis) were 100 to 1 in favor of it? Given 100 to 1 odds in favor of something, I would tend to support it! The statistical procedures in the BayesFactor package calculate these odds ratios for us and thereby give us a clear way of talking about the strength of evidence in favor of one model or another. We can display the odds ratio for our anovaBF() analysis of the three participation groups simply by requesting the output object at the command line:

```
precipBayesOut
```

That command produces the following output:

```
Bayes factor analysis
--------------
[1] precipGrp : 0.1615748 ±0.01%
Against denominator:
Intercept only
---
Bayes factor type: BFlinearModel, JZS
```

The anovaBF() analysis has produced a single value with an uncertainty estimate around it. The value represents the odds ratio in favor of the alternative hypothesis that the groups designated by precipGrp have different means.

Interpreting Bayes Factors

Okay, so now we have these Bayes factors and they are odds ratios comparing one hypothesis to another. Now what? We do not want to choose an arbitrary minimum threshold as the researchers did in days of yore with the *p*-value. As well, we do not want to get into the business of saying things like, "My research has a bigger Bayes factor than your research, so it must be better." To avoid these and other mental traps, let's agree on some essential principles for the interpretation of Bayes factors:

1. Bayes factors are appealing to statisticians and researchers because, in some cases, they can be computed simply and directly from the data (that's why we needed to generate our posterior distributions for the ANOVA in a separate call to the posterior() command). As such, they simply represent one new way of looking at the statistical models and data we have in hand. So you should certainly look at Bayes factors when you can get them, but keep looking at other kinds of results as well (such as 95% HDIs).

2. Every Bayes factor represents a comparison between two statistical models, such as an alternative hypothesis versus a null hypothesis. Whichever model is favored, you can think of the Bayes factor as capturing the amount of "improvement" of the better model over the worse one. Keeping all of that in mind, it is possible that both models that we are comparing may be poor choices, with one of the models somewhat less poor than the other. Choose your research hypothesis and your null hypothesis carefully, lest you compare two things where neither of them is particularly useful.

3. To that end, dedicate some time to thinking about what hypotheses you will compare before you collect your data. Under the frequentist thinking of the null hypothesis significance test (NHST), we often restricted our thoughts to considering how to reject a null hypothesis of no differences of any kind. Bayes factors provide opportunities to compare different alternative hypotheses to each other, instead of only to a null hypothesis.

4. Assuming you are comparing two meaningful and appropriate models, the strength of the Bayes factor is worth interpreting. Kass and Raftery (1995) suggested that any odds ratio weaker than 3:1 is not worth mentioning, odds ratios from 3:1 up to 20:1 are positive evidence for the favored hypothesis, odds ratios from 20:1 up to 150:1 are strong evidence, and odds ratios of more than 150:1 are very strong evidence for the favored hypothesis.

5. The strength of evidence you may need depends upon your research situation. If you are testing a cancer treatment with troubling side effects, you need strong evidence for the efficacy of the treatment. If you are testing a fun technique for improving group dynamics, you can accept modest evidence for the usefulness of your technique. You can think about your Bayes factor in terms of bets and consequences. You might bet on 4:1 odds if you don't have much to lose, but if there's a lot at stake you'd rather have odds of 50:1 or 100:1 in your favor.

Generally speaking, when seeking support for the alternative hypothesis, we want this odds ratio to be as high as possible and powerfully in excess of 1. If you were taking a bet on some uncertain result, wouldn't you want at least 10:1 odds or something like that? In fact, this odds ratio is less than 1, 0.162, which means that the odds favor the null hypothesis in this case (as we would expect given the origins of these data). To make the 0.162 more intuitive, you can invert it like this, $1/0.162 = 6.2$, in other words 6.2:1 odds in favor of the null hypothesis. As a rule of thumb, Kass and Raftery (1995) consider odds between 3:1 and 10:1 to be "positive" evidence for the favored hypothesis, so these data meaningfully support the null hypothesis (also see Jeffreys, 1998, p. 432).

Note the differences between this statement and the results of the traditional frequentist ANOVA. The results from the traditional ANOVA were that we "failed to reject" the null hypothesis, but we did not have anything to say about it beyond this weak statement that we were not able to find any evidence against it. In contrast, the Bayesian procedure gives us a straightforward statement of the odds in favor of the alternative hypothesis (or its inverse, the odds in favor of the null hypothesis).

As a footnote about Harold Jeffreys, in the last line of the anovaBF() output, "Bayes factor type: BFlinearModel, JZS," the JZS acronym stands for Jeffreys–Zellner–Siow prior. Arnold Zellner and Aloysius Siow (1980) of the University of Chicago extended the earlier work of Harold Jeffreys to develop a method of setting prior probabilities when the baseline hypothesis being tested in a multivariate model is a so called "sharp" hypothesis (e.g., a null hypothesis in which the effect size is zero). In practice, it is rarely the case that an effect size is truly zero, so the priors used in this method are in one sense a conservative approach to statistical inference. Where we find strong odds in favor of the "sharp" hypothesis that an effect is completely absent, it is indeed strong evidence against the alternative hypothesis.

FINDING AN EFFECT

In the earlier material, we intentionally configured a situation where the null hypothesis was, in most senses of the word, "true." We literally drew all three of the test samples from the same population. Because much of the frequentist approach to inference is based on having a null hypothesis as a baseline, it was important for you to see the reasoning carried out over a set of data that did not contain an effect—meaning that no differences between group means were anticipated. Keep in mind that if we had repeated the "experiment" 100s or 1,000s of times, we would probably have gotten the null hypothesis test to tell us that there was a meaningful effect there every once in a while. This occurs just because of sampling error, but we know from the procedure we used to create the data that the effect was in fact absent.

Now let's work with some real data that contain a real effect and see what the statistics tell us. In this case, we will use the built-in "chickwts" data set that contains $n = 71$ observations of 6-week-old chicks. At the time of hatching, the chicks were randomly assigned to one of six feed groups. For example, one group ate soybeans while another ate sunflower seeds. Each observation has the feed type in a factor called "feed," while the dependent variable, weight in grams, is in a variable aptly called "weight." I encourage you to use str(), summary(), and barplot() to explore these data before undertaking the analyses shown below. The essential research question is whether the type of feed affects the growth of chicks. If we answer that question affirmatively, it will also be useful to learn which types of feed promote more growth and which types produce less growth.

Let's begin with the standard ANOVA null hypothesis test. Remember that we are interested in whether the variation among means, when scaled properly, exceeds the variance within groups. The ratio of these two quantities produces an F-value. Using the degrees of freedom between groups and degrees of freedom within groups, we position the observed value of F on the appropriate F distribution in order to determine the corresponding p-value (which is the same thing as the area in the tail of the F distribution):

```
data(chickwts)                                    # Probably not needed
chicksOut <- aov(weight ~ feed, data=chickwts)    # Run the ANOVA
summary(chicksOut)                                # Show the ANOVA table
```

The summary() command displays the typical ANOVA table on the R console:

	Df	Sum Sq	Mean Sq	F value	Pr(>F)
feed	5	231129	46226	15.37	5.94e-10 ***
Residuals	65	195556	3009		

Signif. codes: 0 '***' 0.001 '**' 0.01 '*' 0.05 '.' 0.1 ' ' 1

These results show that the p-value on the F-tests has surpassed the traditional 0.05 level of alpha and is therefore statistically significant. Note the use of scientific notation for the p-value: 5.94e-10 is the same thing as saying 0.000000000594. This ANOVA has 5 degrees of freedom between groups and 65 degrees of freedom within groups, so the conventional way of reporting this F-test would be, $F(5,65) = 15.37$, $p < .05$. Actually, in this case many researchers would report $F(5,65) = 15.37$, $p < .001$ to underscore that the result surpassed the most stringent alpha level in typical use. Related to that, you will see that the last line of the output shows how asterisks are used to abbreviate significant levels: three asterisks means the same thing as saying $p < .001$. Clearly we must reject the null hypothesis based on this analysis. As a reminder, the null hypothesis is that all six groups were sampled from the same population such that any variation among means was attributable to sampling error. We can also calculate the eta-squared effect size using the sums-of-squares from the table

231,129/(231,129+19556) = 0.54, suggesting that feed type explained about 54% of the variance in weight. Armed with a rejection of the null hypothesis, we should go on and look more carefully at the group means, and possibly conduct some follow-up tests to precisely examine the mean differences between different pairs of groups, what statisticians call **post hoc testing.**

Before we go there, however, let's conduct the Bayes factor ANOVA and see what we learn from that:

```
# Calc Bayes Factors
chicksBayesOut <- anovaBF(weight ~ feed, data=chickwts)
# Run mcmc iterations
mcmcOut2 <- posterior(chicksBayesOut,iterations=10000)
boxplot(as.matrix(mcmcOut2[,2:7]))    # Boxplot the posteriors for the groups
summary(mcmcOut2)                     # Show the HDIs
```

You can review the boxplot of the posterior distributions in Figure 6.7. Remember that these represent deviations from the grand mean, so some groups will be on the positive side and some groups will be on the negative side. The zero point on the Y-axis therefore represents the position of the grand mean.

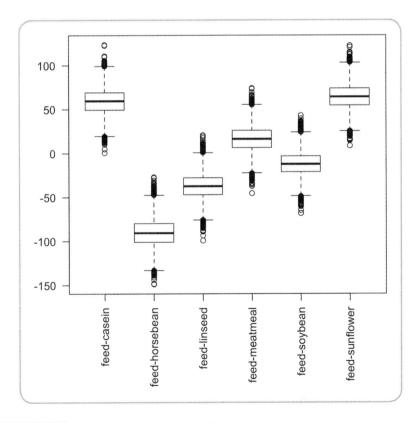

FIGURE 6.7. Box plot of the posterior distributions for the chickwts feed groups.

There are two groups, casein on the far left and sunflower on the far right, that are similar to each other and quite obviously superior to the other groups. The only other group with a mean that is above the grand mean is the meatmeal group, but its distribution overlaps with the linseed and soybean groups quite substantially. The loser is horsebean, which also overlaps with linseed and to a small degree with soybean. Note that I am paying attention to the whiskers for each box: for informal purposes you can consider these to be roughly the same as the boundaries of the 95% HDI for each distribution. But for a detailed view of the HDIs we should really consult the second half of the console output from the summary() command above:

2. Quantiles for each variable:

	2.5%	25%	50%	75%	97.5%
mu	246.1102	254.7790	259.236	263.884	272.637
feed-sunflower	35.5407	54.4183	64.198	74.220	93.330
feed-casein	30.4805	49.8965	59.270	69.119	87.816
feed-meatmeal	-12.6600	6.3583	16.712	26.414	45.732
feed-soybean	-38.4556	-21.4666	-12.178	-3.364	14.310
feed-linseed	-65.9157	-47.2791	-37.497	-28.080	-9.537
feed-horsebean	-122.0330	-101.1611	-90.387	-79.524	-58.058

For the sake of clarity I have reordered the rows of the output so that you can see the pattern more clearly. The median for mu, the grand mean, is at 259.2 grams, so all of the group deviations are in reference to that. Sunflower and casein have HDIs that are quite positive, with median deviations of 64.2 and 59.2, respectively. The HDIs for these two types of feed largely overlap, so the difference between the two is negligible. Meatmeal is next, and the top of its HDI overlaps with the bottom of the HDIs for sunflower and casein, so you might want to explore those pairwise comparisons further with post hoc testing. The HDI for soybean straddles 0 and the top end of it overlaps very slightly with the bottom end of meatmeal. Finally, linseed and horsebean are both solidly on the negative side of the ledger and neither of these have any degree of overlap with the top two. Based on these HDIs alone, we can conclude that we have two strong winners, sunflower and casein, as well as two clear losers, linseed and horsebean. By the way, what is horsebean?

To conduct any pairwise "post hoc" tests on meatmeal or soybean to sort out whether they are credibly different from other groups, you can easily use the BESTmcmc() function to run a Bayesian *t*-test comparing any two groups. For example, the following command would test whether meatmeal and sunflower are credibly different from one another:

```
plot(BESTmcmc(chickwts[chickwts$feed=="sunflower",1],
    chickwts[chickwts$feed=="meatmeal",1]))
```

We already have a strong sense of the results of this ANOVA from the conventional null hypothesis significance test (NHST) and from the Bayesian

HDIs of the posterior distribution. We have one additional piece of evidence at our disposal, specifically the Bayes factor. As before, just request the output by naming the object from the anovaBF() analysis on the R console:

chicksBayesOut

Which provides the following output:

```
Bayes factor analysis
--------------
[1] feed : 14067867 ±0%
Against denominator:
Intercept only
---
Bayes factor type: BFlinearModel, JZS
```

This analysis shows odds of 14067867:1 in favor of the alternative hypothesis (with zero margin of error!). According to the rules of thumb provided by Kass and Raftery (1995), any odds ratio in excess of 150:1 is considered very strong evidence. This result confirms our previous evidence suggesting support for an alternative hypothesis of credible differences among these group means.

CONCLUSION

ANOVA is one of the most prevalently used statistical tools in the sciences and is particularly useful in those circumstances where researchers run experiments to compare different groups. You may recall from my description of the most recently used data that the chicks were randomly assigned to feed groups upon hatching. The use of random assignment of cases to different, but comparable conditions (as well as a few other key features), is what makes an experiment an experiment. The fact that an experiment contains multiple groups, where each group contains observations of a dependent variable, makes ANOVA one of the preferable tools. In particular, when you have a metric variable as the dependent variable and one or more grouping variables as the independent variable(s), an ANOVA analysis provides a straightforward method of making comparisons among two or more group means. When I was learning statistics I wondered whether it was correct to use ANOVA in situations where a *t*-test would do, specifically when there were exactly two groups to compare. The answer is that either ANOVA or a *t*-test will do in that circumstance, but for more than two groups ANOVA is an excellent choice because it can evaluate, in a single test, whether there are *any* differences among the groups being compared.

We reviewed several strategies to conduct the ANOVA test. The traditional frequentist method tests the null hypothesis that the between-groups variance and the within-groups variances are proportional. The *F*-ratio that expresses this proportionality shows the extent to which the group means diverge with one another. When we calculate the observed F-ratio that was calculated from the respective variances, we position that value on a (probability density) distribution of F that is appropriate for the degrees of freedom (df) in our analysis. The F distribution is a family of distributions whose specific shape is dependent upon the df between groups and the df within groups. For example, our ANOVA of chickwts data was evaluated on F(5,65) degrees of freedom, meaning five df for between groups and 65 df within groups. When we evaluate an observed value of F, we determine the corresponding p-value, which equates to the area under the curve beyond the observed value of F. That's why we state the null hypothesis p-value as the probability of observing an F-value this high or higher, under the assumption that the null hypothesis is true.

That is a mouthful, which is one of the reasons that HDIs are so appealing. The Bayesian approach to ANOVA, provided in this case by the BayesFactor package, provided us with the ability to sample from posterior distributions of population parameters. In addition to a population mean, mu, which we also referred to as a grand mean, we obtained distributions for the deviations of group means from the grand mean. Using graphical and tabular methods, we examined the HDI for each group and ascertained the extent that one group's HDI overlapped with another group's HDI. When the HDIs for two groups do not overlap, there is credible evidence for different group means in those two groups. To the extent that two groups' HDI overlap there is evidence of no credible difference between the means. I suggested that you could use the BESTmcmc() Bayesian t-test function to conduct a pairwise test in those circumstances where this process of comparing HDIs for overlap did not yield a conclusive result. (Bonus fact: the BayesFactor package also contains a t-test procedure: the command is called ttestBF.)

The BayesFactor output, true to its name, also provided us with a Bayes factor that compares two hypotheses using an odds ratio. One hypothesis is that all group means are equivalent to mu, the grand mean. This corresponds to the null hypothesis for traditional ANOVA. Another hypothesis is that the independent variables, also sometimes called "factors," account for deviations of the group means from the grand mean. This is comparable to the alternative hypothesis for traditional ANOVA. When we create a ratio of the likelihoods of these two different hypotheses, we get an odds ratio that can be interpreted as the odds in favor of one or the other hypothesis. Generally, though not always, we hope to find evidence in favor of the alternative hypothesis, with odds in excess of 3:1, 20:1, or 150:1, depending upon the research situation.

All of these tools will be useful when you want to compare groups on a metric variable. You are likely to observe the traditional null hypothesis ANOVA in many journal articles, past and present. When you see the F-test and the corresponding p-value reported in these contexts, you will know how to make sense of whether the statistical results were significant or not significant. Rather than stopping there, however, it is valuable to obtain an effect size, which in the case of ANOVA is the eta-squared value, corresponding to the proportion of variance in the dependent variable accounted for by the independent variables. In your own data analysis, you could also use the Bayesian approach for ANOVA. Ideally, this yields highest density intervals (HDIs) that you can use to look for credible differences among group means. If there is ambiguity about the separation between any pair of groups, you can follow up with a Bayesian *t*-test. In addition, you can examine the Bayes factor to review the weight of evidence in favor of the alternative hypothesis. When reporting the results of your own research to others or in publications, it might make sense to use all of these types of information, fit them together as a mosaic of evidence, and then provide your views on the strength of the evidence when taken as a whole.

EXERCISES

1. The data sets package (installed in R by default) contains a data set called InsectSprays that shows the results of an experiment with six different kinds of insecticide. For each kind of insecticide, *n* = 12 observations were conducted. Each observation represented the count of insects killed by the spray. In this experiment, what is the dependent variable (outcome) and what is the independent variable? What is the total number of observations?

2. After running the aov() procedure on the InsectSprays data set, the "Mean Sq" for spray is 533.8 and the "Mean Sq" for Residuals is 15.4. Which one of these is the between-groups variance and which one is the within-groups variance? Explain your answers briefly in your own words.

3. Based on the information in question 2 and your response to that question, calculate an *F*-ratio by hand or using a calculator. Given everything you have earned about *F*-ratios, what do you think of this one? Hint: If you had all the information you needed for a Null Hypothesis Significance Test, would you reject the null? Why or why not?

4. Continuing with the InsectSprays example, there are six groups where each one has *n* = 12 observations. Calculate the degrees of freedom between groups and the degrees of freedom within groups. Explain why the sum of these two values adds up to one less than the total number of observations in the data set.

5. Use R or R-Studio to run the aov() command on the InsectSprays data set. You will have to specify the model correctly using the "~" character to separate the dependent variable from the independent variable. Place the results of the aov() command into a new object called insectResults. Run the summary() command on insectResults and interpret the results briefly in your own words. As a matter of good practice, you should state the null hypothesis, the alternative hypothesis, and what the results of the null hypothesis significance test lead you to conclude.

6. Load the BayesFactor package and run the anovaBF() command on the InsectSprays data set. You will have to specify the model correctly using the "~" character to separate the dependent variable from the independent variable. Produce posterior distributions with the posterior() command and display the resulting HDIs. Interpret the results briefly in your own words, including an interpretation of the BayesFactor produced by

the grouping variable. As a matter of good practice, you should state the two hypotheses that are being compared. Using the rules of thumb offered by Kass and Raftery (1995), what is the strength of this result?

7. In situations where the alternative hypothesis for an ANOVA is supported and there are more than two groups, it is possible to do post-hoc testing to uncover which pairs of groups are substantially different from one another. Using the InsectSprays data, conduct a *t*-test to compare groups C and F (preferably a Bayesian *t*-test). Interpret the results of this *t*-test.

8. Repeat Exercises 5, 6, 7, but this time using the built-in PlantGrowth data set. Create a written report of the results (preferably with graphics) that would be of high enough quality to be used in a professional presentation.

9. Repeat Exercise 8, with the built-in attenu earthquake data set. The research question is to find out whether the events differed from one another in their mean acceleration. Hint: You will need to coerce the numeric "event" variable into a factor.

## CHAPTER 7

Associations between Variables

One of the fundamental survival skills of the human species, and arguably a capability that has promoted the development of civilization and technology, is our ability to see patterns in events. If we could somehow look back 100,000 years and see our long lost ancestor Og building a fire, we might notice that Og put more wood on the fire when he wanted more heat, and less wood on the fire when he wanted less heat. When Og recognized the connection between more wood and more heat, he understood one of the most fundamental probabilistic patterns in nature: the association.

Thanks to his big brain, Og learned that there was only an imperfect correspondence between the amount of wood and the amount of heat. For example, when green wood burns it does not burn as strongly as dry wood. This observation underscores the fact that most associations in nature are imperfect. Some logs will burn hotter and more quickly while others may not catch fire at all. The connection between the amount of fuel on the fire and the heat of the fire is inexact. The implication here is that associations vary in their "strength"—some associations are strong, almost mechanical in the way they seem to link phenomena. Other associations are weak and difficult to spot, but may nonetheless be important and meaningful.

Statistical measures of association were among the first multivariate statistics developed. One of the most famous measures of association—the Pearson product–moment correlation (PPMC)—was developed by the late 19th-century statistician Karl Pearson (1857–1956). Pearson's brilliant historical record as a mathematical statistician is somewhat tainted by his involvements with Francis Galton and the eugenics movement, but if you use contemporary statistics you owe something of a debt of gratitude to Pearson for the many techniques he developed. By the way, our Guinness guy, William Sealy Gosset, whom we also know as "Student," was a colleague of Pearson and worked in his lab (Plackett & Barnard, 1990, p. 30).

From a technical standpoint, the PPMC builds upon a very powerful concept known as **covariance.** We are familiar with variance, from earlier in this book, as a measure of the variability of a numeric variable. Covariance builds on this concept by considering the variances of two variables in combination. If we sampled a variety of fires started by our ancestor Og, and we measured the amount of fuel and the resulting heat of each fire, we could *partition* the variance of the heat and fuel variables into two components: a shared component and an independent component. Some amount of the variance in heat will be in common with the variance in fuel. The remaining amount of variance will not be in common. In essence, the ratio of common variance—*covariance*—versus independent variance is the correlation between the two variables.

Let's explore this numerically and visually with R. We can begin with a situation where there ought to be no association between two variables. The following code generates two random samples of data, with $n = 24$ observations each, roughly normally distributed, with means pretty close to 0 and standard deviations pretty close to 1:

```
set.seed(12345)
wood <- rnorm(24)
heat <- rnorm(24)
mean(wood)
mean(heat)
sd(wood)
sd(heat)
```

You should run those commands yourself and check out the results. The first command, set.seed(), controls the start of the randomization so that you and I will get the same results. If you leave this out or change the seed number you will get different results. The means are close to zero and the standard deviations are close to one, signifying that there are both positive and negative values within each vector. I know it seems kind of weird to have "negative wood" and "negative heat" but just go with it for now by considering these negative values to simply be the lowest amounts of wood and heat. The next thing to do is to look at a plot of these variables, using the plot(wood, heat) command as shown in Figure 7.1.

You might notice in Figure 7.1 that the ranges of the wood and heat variables are roughly –2 to +2 and that there are quite a number of points near the respective means of the two variables, right at the center of the plot. Think about why this is typical: the normal distribution has many of its points clustered within one standard deviation of the mean, and almost all of its points within plus or minus two standard deviations. The idea of a "standard normal" distribution indicates that the mean of a distribution is 0 and its standard deviation is 1. In the case of our wood and heat variables, the only reason there is a slight deviation from that is because of randomness and the small size of these samples.

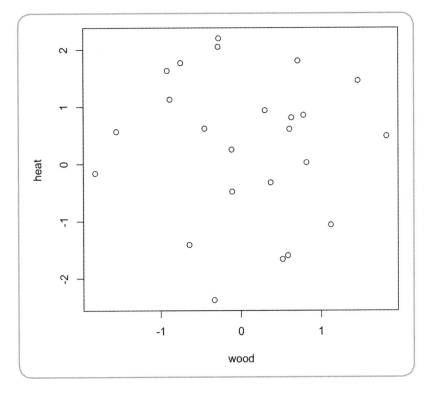

FIGURE 7.1. Scatterplot of two randomly generated, standard normal variables.

Nonetheless, for the purposes of this example we can loosely call both wood and heat standard normal variables. This also means that each observation of wood and heat is, in effect, a so-called *z-score;* in other words, each of these scores is a deviation from the mean, calibrated in standard deviations. You can verify this yourself by plotting values of either variable versus its own deviations from the mean, with the following command, the results of which appear in Figure 7.2:

```
plot(wood,(wood-mean(wood)))
```

The fact that Figure 7.2 shows a straight line, with both X- and Y-axes calibrated identically, shows that each value of wood and its associated deviation from the mean are essentially the same thing. Given that the mean of wood is close to 0, this makes total sense. We can get a flavor of how much these two variables covary by calculating the *products* of their respective deviations from the mean. Think about how this will work conceptually: if each member of a pair of deviation scores is positive, the product will be a positive number. So to the extent that a particular fire has both lots of wood and high heat, there will be lots of positive cross-products. Likewise, if each member of a pair is a

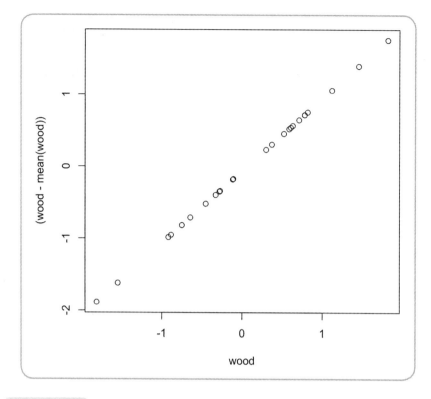

FIGURE 7.2. Scatterplot of a standard normal variable versus deviations from its own mean.

negative number, the cross-product will also be a positive number. So to the extent that a particular fire has very little wood and very low heat, there will also be lots of positive cross-products. In some cases, the variables may "pull" in opposite directions: one variable will have a positive z-score while the other has a negative z-score. This will result in a negative cross-product that will diminish the overall covariance. If we look at the mean of these cross-products across all pairs of data points we should be able to get an idea of how much they covary. Let's try it:

```
cpWH <- wood * heat
mean(cpWH)
hist(cpWH)
```

The first command creates the cross-products of the z-scores of wood and heat. The second command shows the mean cross-product. A histogram summarizing all 24 of the cross-products appears in Figure 7.3.

Remember that the two variables we created were randomly generated, so we do not expect to find any association between them. The histogram backs this up: about 18 of the cross-products are right near 0 (the four central bars in

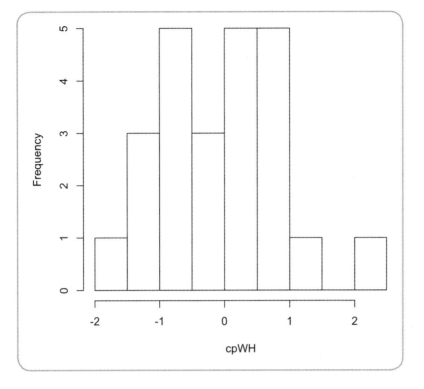

FIGURE 7.3. Histogram of the crossproducts of wood and heat.

the histogram), which means that for more than half the pairs of deviations in the data set one or the other member of the pair was near 0. In fact, there are only about two-cross products that are positive and a few negative ones basically cancel out these two positive ones. In the code above, we also called for mean(cpWH), in other words, the *average* cross-product between pairs of variables. This turns out to be about −0.05, which, as we will learn, is a very tiny value in the world of the PPMC. We can essentially consider it to be equivalent to 0—meaning no association between the variables.

Next, let's "cook" (get it?) our data to force a positive correlation between wood and heat. This is pretty easy to do for this particular example because wood and heat *are on precisely the same scale.* Because we know that they are both, roughly speaking, standard normal variables, we can do the following transformation and expect it to work well. Do not expect this to work if the means and/or standard deviations of the two variables are different from each other:

```
# Make a new, fake version of heat that will correlate with wood
newHeat <- wood/1.41 + heat/1.41    # Make a mixture of the two old variables
mean(newHeat)    # What's the mean of our new heat variable?
sd(newHeat)        # What's the SD of our new heat variable?
```

The first command creates a new, fake version of our heat variable by taking half of its influence from the wood variable and half of its influence from the existing heat variable. (Note that the 1.41 in denominator is the square root of 2, which is what we need mathematically to have the standard deviation of newHeat come out close to one.) In other words, we are forcing our variable newHeat to have some covariance with wood by intermixing the variability in wood with the variability in our original heat variable. The mean() and sd() commands confirm that the newHeat variable is still roughly standard normal. The mean is quite close to zero and the standard deviation is close to one. For our purposes, we can still consider these to be deviation scores and so we can do the same cross-product trick that we did before:

```
cpWnewH <- wood * newHeat
hist(cpWnewH)
mean(cpWnewH)
```

Now we have a new vector, cpWnewH that contains cross-products of wood and the newHeat variable that we intentionally constructed to have a lot of influence from wood. What do you expect to find in this vector? Based on our earlier reasoning, there should now be lots of positive cross-products, where wood and newHeat are commonly both positive or both negative in many of the pairs. As a result, we should also see a positive mean of the cross-products that is considerably higher than the −0.05 that we saw earlier. A histogram of the cross-products appears in Figure 7.4.

The histogram in Figure 7.4 shows a notable difference from the one in Figure 7.3. In this case all but nine of the cross-products are above 0 and there is a substantial positive tail of values near 3 that are not offset by any corresponding negative values. The mean of the cross-products confirms this: I got a value of 0.52549589; with rounding, let's call it 0.53.

We can see this visually in a scatter plot as well. The plot(wood,newHeat) command will produce the graph we need. In the resulting Figure 7.5, notice how the cloud of points is somewhat "cigar shaped"—that is, thin at the lower and upper ends and thick in the middle—with a clear linear trend. If you were to draw a line through these points, you would start at the origin on the lower left and you would end up at the upper right corner, and most of the points along the way would fall pretty close to the line.

So now we have a nice intuitive and graphical sense of the meaning of the PPMC. Remember that the calculations we did with cross-products will only work if you have two variables that are both standard scores (aka z-scores), but the basic principles and methods are the same for nonstandardized variables. The association between the two variables is the "moment," that is, the mean, of the products of the deviations of two variables, mathematically adjusted for the amount of variability in each variable. Armed with this knowledge, let's now have R calculate the actual correlation with its built-in procedure for the PPMC:

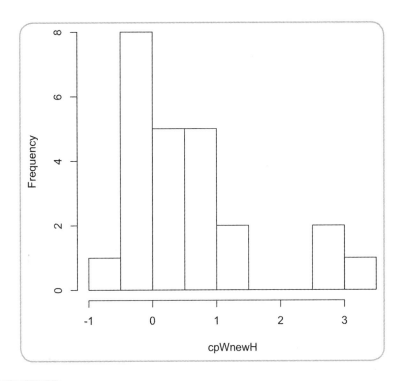

FIGURE 7.4. Histogram of the cross-products of wood and newHeat–a new variable with forced correlation to wood.

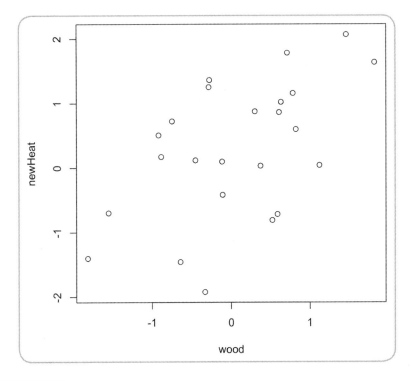

FIGURE 7.5. Scatterplot of wood and newHeat—a new variable with forced correlation to wood.

cor(wood,newHeat)
[1] 0.5464432

The cor() command produces a PPMC based on the two variables you supply. Notice that this value, about 0.55, is slightly larger than the mean product moment of 0.53 that we calculated earlier. This small discrepancy is a result of the fact that the standard deviations of the wood and newHeat variables were not exactly one, as they should have been if these were truly standard scores (also known as z-scores). If you increased the sample size, say up to $n = 2,400$, you would get vectors of data that more closely approximated the standard normal distribution and the two methods of calculating the correlation would come out even more closely. Try it! If you rerun the code above with larger samples, you may also find the scatterplots quite instructive.

Formula for Pearson's Correlation

Pearson's product–moment correlation (PPMC; usually abbreviated as r) is the single most commonly used measure of association between variables. As the name says, it is a "moment" (which is like saying "mean") of a set of products. At the beginning of this chapter we used standard scores to explore the concept of a product moment. Remember that r ranges from –1 up to 1, signifying that r itself is on a kind of standardized scale. In the general case, in order to accommodate variables with different amounts of variance, we need a method of calculating r that takes into account the variability of each of the two variables. The following formula represents one strategy for calculating r:

$$r = \frac{\Sigma(x - \bar{x})(y - \bar{y})}{\sqrt{\Sigma(x - \bar{x})^2 \, \Sigma(y - \bar{y})^2}}$$

The numerator contains the sum of the cross-products of the deviations from the respective means of the two variables. The denominator contains separate calculations for the sum of squares for each of the two variables. These sums of squares are then multiplied together and then the square root of that product is taken. For all three of the summation symbols (the Greek sigma, which looks like a capital E), we are summing across all the observations in the sample. You may be wondering that if this formula represents a "moment," would we not have to divide by the sample size at some point to get a mean? The answer is yes, but because we divide by the sample size in both the numerator and the denominator, the sample size cancels itself out of the calculation.

You can test out the logic of this formula by imagining a sample with just two observations, $\{x = -1, y = -1\}$ and $\{x = 1, y = 1\}$. With this simple data set, the mean of x is 0 and the mean of y is 0, simplifying both the numerator and the denominator. Try it out and see what you get.

One thing to remember about notation: Karl Pearson was working with Francis Galton when he cooked up the mathematics for the correlation coefficient, and at the time, Galton was referring to the quantity as "reversion" and "regression" because he was interested in the idea of "regression to the mean." So Pearson called the correlation value "r" as a result. If you look up formulas for r you will often find the Greek letter "rho," which looks a lot like our letter p. Rho is used to refer to the "true population value" of the correlation, whereas "r" is used to refer to a correlation calculated from a sample. In the material that follows, I will frequently use the small letter r to refer to a correlation estimated from a sample (as opposed to uppercase R which is the name of our statistical software) and the word "rho" to refer to the population value of the correlation coefficient.

INFERENTIAL REASONING ABOUT CORRELATION

In the material above, I "cooked up" (there I go again!) a pair of variables, called wood and heat, which, in the first place, had a very low correlation between them, because they were randomly generated. Then I created "newHeat" based on intermixing the data from wood and heat. This created a stronger correlation between the two fake variables, and we were able to see the value of r rise from about $r = -.05$ to $r = .55$. Note that r ranges from -1.0 to $+1.0$ and that any values close to 0 indicate weak correlations, whereas values near -1.0 or $+1.0$ indicate strong correlations. When a correlation has a minus sign on it, this means that there is an inverse association between the two variables. For example, putting water on a fire cools it off: the more water, the lower the heat.

We know from our explorations of sampling earlier in the book that each time we draw a sample we get a slightly different configuration of data. Over the long haul, summaries of these samples show that most of the statistics are close to the population value but it is also the case that some sampling error usually occurs. This is just as true with correlations as it is with other types of statistics: A correlation of $r = .55$ in a sample *does not* indicate that rho equals that same value: depending upon the nature of the sample, the population rho may be quite a bit different from that $r = .55$ value. Given two variables in a sample with a nonzero correlation between them, we would like to know what this implies about the true population value of rho. How much uncertainty is there around our observed value of r and how much can we trust that it is truly nonzero in the population? We can use inferential reasoning to think about these questions and give us some guidance.

Let's make an informal model of the population and random sampling, as we have done before, to look at the range of possible outcomes. While before we were sampling from a single variable, what we need to do in this case is to sample pairs of variables at random from a data set. Let's start by creating a fake "population" of values for each of the two variables and placing it in a data frame:

```
set.seed(12345)                    # Start with a random number seed
wood <- rnorm(2400)                # Make two vectors of N=2400
heat <- rnorm(2400)
fireDF <- data.frame(wood, heat)   # Put them in a data frame
nrow(fireDF)                       # Verifying 2400 rows of two variables
fireDF[sample(nrow(fireDF), 24), ] # Generates one sample of n=24
```

You should verify for yourself that the correlation between wood and heat in our fake population is near 0. In the final line of code above you may be able to spot that we are using the sample() function in a new way. We are using it to choose, at random, a list of 24 row numbers from anywhere in the sequence of 1 to 2,400. Each row that gets chosen by sample() will end up in a new "mini" data frame. This new data frame will contain a sample from the "population" data frame. We can then calculate a Pearson's r correlation from that sampled set of data, like this:

```
cor(fireDF[sample(nrow(fireDF), 24), ])
```

```
                wood        heat
wood      1.00000000  -0.01249151
heat     -0.01249151   1.00000000
```

You will notice that the output from this command is a correlation matrix, rather than a single correlation. That's a bit of a nuisance, because we only need the value in the lower left corner. FYI, the 1.0 values in the diagonal are the correlations between each variable and itself, which, not unexpectedly are 1.0, the perfect positive correlation. You may also notice that the correlation value $r = -.0125$ appears in two places, once on the upper right and once on the lower left. Of course, the correlation between wood and heat is the same as the correlation between heat and wood. So we only want to take one element from the matrix like this:

```
cor(fireDF[sample(nrow(fireDF), 24), ])[1,2]
```

The notation [1,2] at the end gives us the upper-right (row = 1, col = 2) value in the matrix. We can now replicate this process a few thousand times and look at the results we get:

```
corDist <- replicate(5000,cor(fireDF[sample(nrow(fireDF), 24), ])[1,2])
hist(corDist)
mean(corDist)
```

I used replicate() to generate 5,000 repetitions of running a correlation matrix on a sample of $n = 24$ pairs of data points from the larger dataframe. Then I requested a histogram (see Figure 7.6) of the resulting sampling distribution

Reading a Correlation Matrix

When we provided a data frame to the cor() procedure, instead of getting one correlation coefficient, we got a square table with lots of numbers. Let's generate a correlation matrix with more than two variables and make sense out of the result:

```
cor(iris[,1:4])
```

	Sepal.Length	Sepal.Width	Petal.Length	Petal.Width
Sepal.Length	1.0000000	-0.1175698	0.8717538	0.8179411
Sepal.Width	-0.1175698	1.0000000	-0.4284401	-0.3661259
Petal.Length	0.8717538	-0.4284401	1.0000000	0.9628654
Petal.Width	0.8179411	-0.3661259	0.9628654	1.0000000

I used the cor() command to generate a correlation matrix for the first four columns in the built-in iris data set, which contains measurements of $n = 150$ iris flowers. The resulting table/matrix is square, meaning that it has the same number of rows as it does columns. The diagonal contains all values of 1, indicating the perfect correlation that always exists between a variable and itself. There are two triangles of correlation data, one above the diagonal and one below it. The two triangles are transposed versions of each other: they contain the same information, so you really only need to look at the lower triangle.

Notice that each correlation is given with seven decimal places of precision. That's more than enough in most social science research, because the instruments we use (such as surveys) don't usually measure things so precisely. When a correlation matrix is reported in a journal article or other reports, authors will often leave off the leading 0 on a correlation coefficient, because it is understood by readers that a correlation always falls in the range of -1 to $+1$. Every correlation starts with a 0, so we don't need to keep repeating it. Although scientific reporting styles vary, it would not be surprising for a journal article to contain a correlation matrix that looked more like this:

	1	2	3
1. Sepal.Length			
2. Sepal.Width	-.12		
3. Petal.Length	.87	-.43	
4. Petal.Width	.81	-.37	.96

The diagonal is completely absent, the names of the variables only appear once, and we have dropped the leading zero. I actually find this much easier to read than the full matrix. I quickly note that Sepal.Width has a negative relationship with the other variables. I tend to dismiss the $-.12$ as negligible and focus my attention on the larger correlations, starting with the $-.37$. My eye also falls on the .96 correlation between

(continued)

Petal.Length and Petal.Width—an extremely high value, suggesting the possibility that these two variables might in some senses be redundant with each other.

Finally, in many cases researchers also report the results of a null hypothesis significance test on each correlation coefficient. A coefficient that is marked with one asterisk is significant at $p < .05$, two asterisks at $p < .01$, and three asterisks at $p < .001$. The basic cor.test() function that we use in this chapter does not work on a whole matrix of correlations. If you want to generate null hypothesis tests for a whole matrix of correlations, try the corr.test() function in the "psych" package.

of correlations. Finally, I calculated the mean of the distribution. Before you review the histogram, remember that we are working here with the two uncorrelated variables: wood and heat. The rnorm(2400) commands shown previously created a fake population of random data for us where the two variables should be unrelated to one another.

Pretty striking! Perfectly centered on 0, as we would expect from two randomly generated variables. In fact, the mean is about $r = -.01$. Quite normal in appearance, with light, symmetric tails. Most interesting of all, there are quite a few samples in the tails with values near or in excess of 0.40. These values show the dangers of sampling and the importance of inferential statistics. From small

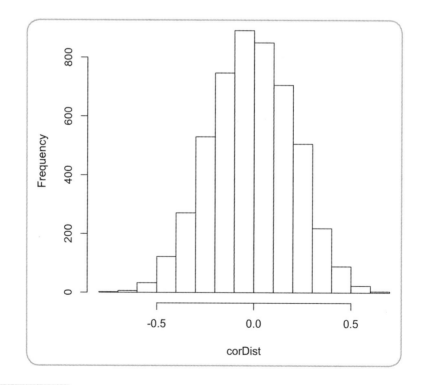

FIGURE 7.6. Histogram of 5,000 correlations from samples of $n = 24$.

samples it is relatively easy to get values of r that look as if they are showing a sizable positive or negative correlation, *even from a population where the variables are uncorrelated.* In fact, let's look at the 0.025 and 0.975 quantiles of this distribution so we can get a feel for what constitutes an extreme value:

```
quantile(corDist, probs=c(0.025,0.975))
       2.5%          97.5%
-0.4234089     0.3976422
```

Yikes! Anything above 0.40 or below −0.42 would be quite rare, but this also shows that we could easily get values of about $r = .25$ or $r = −.25$ because those would fall in the central area of the sampling distribution. Now let's repeat the exercise, but this time we will create a new version of the heat variable that intermixes some influence from wood:

```
newHeat <- wood/1.41 + heat/1.41        # Mix the two source variables
newfireDF <- data.frame(wood, newHeat)  # Add the new one to a dataframe
newcorDist <- replicate(5000, cor(newfireDF[sample(nrow(newfireDF),24),])
      [1,2])
hist(newcorDist)                        # Show a histogram of the r values
mean(newcorDist)                        # What's the mean correlation
```

The histogram in Figure 7.7 tells a different story from the one in Figure 7.6. This histogram centers at about 0.70 with a long tail headed down to about 0.2 (apparently no values in the negative range) and a lesser tail leading up toward 1.0. The actual mean of the distribution calculated by the mean() function is 0.7013. The asymmetric appearance is because the correlation coefficient r cannot go higher than 1.0, so there is a ceiling effect in the upper part of the distribution. We should look at the 0.025 and 0.975 quantiles too:

```
quantile(newcorDist,probs=c(0.025,0.975))
```

This command reveals that the 0.025 threshold is at $r = .43$ and the 0.975 threshold is at $r = .87$. So 95% of all the correlations we sampled from the original population fall in this range. You can think of this range as a kind of very informal 95% confidence interval around the point estimate of $r = .70$. If you are curious, and I hope you are, you can run cor(wood,newHeat) to find the actual calculated value of the correlation coefficient rho (treating our $n = 2,400$ observations as a complete population). You will find that it is very close to the center of the empirical sampling distribution we constructed. One last neat fact: If you square the value $r = .70$ you get the result 0.49. Remember that when we made newHeat, half the influence came from the variable "heat" and half came from the variable "wood" with which we were trying to create a positive correlation. The square of the correlation coefficient yields a value of "r-squared" which you can interpret as the proportion of variance the two variables have

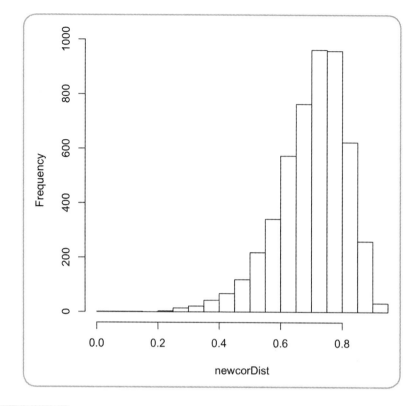

FIGURE 7.7. Histogram of 5,000 correlations from samples of $n = 24$, where rho = .71.

in common. In this case we built a variable "newHeat" that was designed to have half of its variance in common with "wood" and that is reflected in the r-squared value equal to approximately one-half.

Don't forget that this is an informal demonstration only. In the material below we will calculate a real confidence interval for the correlation coefficient. This informal demonstration helped to illustrate graphically how much a correlation coefficient can vary when calculated from a sample of data. Although we get close to the population value of rho in most cases, there are also numerous samples that produce values of r that are quite different from rho.

NULL HYPOTHESIS TESTING ON THE CORRELATION

The previous section gave us a chance to think about sampling distributions of correlations and we ended up in a nice place where we could actually build our own simulated confidence interval around a sample-based estimate of the correlation coefficient r. In keeping with the standard approach to the null hypothesis significance test, the procedure for testing the significance of the correlation coefficient assumes a null hypothesis of rho = 0.

R provides a simple procedure for the null hypothesis test on a correlation called cor.test():

```
set.seed(12345)
wood <- rnorm(24)
heat <- rnorm(24)
cor.test(wood,heat)
```

The cor.test() command above produces the following output:

```
Pearson's product-moment correlation
data: wood and heat
t = -0.2951, df = 22, p-value = 0.7707
alternative hypothesis: true correlation is not equal to 0
95 percent confidence interval:
-0.4546764 0.3494514
sample estimates:
cor
-0.06279774
```

Note that in the code above I have returned to using a small random sample of $n = 24$ observations. I did this on purpose so that you would get a flavor for the significance test and the confidence intervals that are consistent with our first example near the beginning of this chapter. Don't get confused by the fact that the histograms in Figure 7.6 and Figure 7.7 represent sampling distributions for samples of $n = 24$ observations and not plots of raw data.

The output above has three sections: The first three lines are the conventional null hypothesis test with an assumption of rho = 0. The test statistic is a t-test on a transformed version of the correlation coefficient. The test yields a very tiny t-value of $t = -0.2951$. Remember that any t-value with an absolute value less than about 2 is unlikely to be significant. The phrase df=22 refers to the degrees of freedom. As we learned from the ANOVA chapter, degrees of freedom is a mathematical concept that statisticians use to keep track of how many elements are free to vary in a statistical system. In this case, starting with $n = 24$ observations, one degree of freedom was lost for the calculation of the mean in each of the two samples. The value of $t = -0.2951$ is well inside the central region of the t distribution for df=22, so the corresponding probability value of 0.7707 is quite sensible. One way of thinking about this p-value is to say that there is a 0.7707 chance of observing an absolute value of t this high or higher under the assumption that the population value of rho = 0. Using the conventional $p < .05$ threshold for alpha to evaluate this result, we have *failed to reject* the null hypothesis of rho = 0.

The cor.test() procedure also provided a 95% confidence interval around the point estimate of $r = -.06$. If you compare the confidence interval displayed here to the one we constructed using the empirical sampling distribution displayed in Figure 7.6, you will find that they are quite similar. The technical definition of the confidence interval would be this: if we repeated this sampling process many times and each time constructed a confidence interval around the calculated value of r, about 95% of those constructed intervals would contain the true population value, rho. In reporting this confidence interval in a journal article, we would simply say that the 95% confidence interval for rho ranged from −0.45 to 0.35. This is a very wide range indeed and it reflects the fact that $n = 24$ is a small sample size from which to make an inference. Importantly, the confidence interval straddles 0, a result that concurs with the results of the significance test.

In contrast to this result, let's reconstruct a fake variable that is half-random, uncorrelated data from "heat" and half influence directly taken from "wood." This will force a correlation to exist between the two variables. We will do the calculation on the fly this time rather than bothering to create a new variable again:

```
cor.test(wood,(wood/1.41 + heat/1.41))
```

That call to cor.test() produces the following output:

```
Pearson's product-moment correlation
data: wood and (wood/1.41 + heat/1.41)
t = 3.0604, df = 22, p-value = 0.005731
alternative hypothesis: true correlation is not equal to 0
95 percent confidence interval:
0.1834949 0.7782808
sample estimates:
cor
0.5464432
```

This time we have detected statistical significance: we have a t-value of $t = 3.06$ with an associated p-value of $p = 0.0057$. This value of p is considerably lower than our conventional alpha level of $p < .05$, so we can safely say that we *reject the null hypothesis* of rho $= 0$. Again here I encourage you to compare the confidence interval of {0.18, 0.79} to the empirical values we obtained from the sampling exercise displayed in Figure 7.7. Naturally, they do not match up exactly because of random sampling variations, but the range is similar.

Before we move on, let's run one example with some real data, just to get the feel for how the null hypothesis test works in a more typical research situation. For this example, we will use the classic iris data set collected by Edgar Anderson (1935). The iris data set is built in to the basic installation of R, so you should not normally have to run the command data(iris). The data set contains four measurements of characteristics from a sample of $n = 150$ iris plants. We will test whether there is a significant (i.e., nonzero) correlation between the width of a "sepal" (a leafy part under the flower) and the width of a petal, using the following command:

```
cor.test(iris[,"Sepal.Width"],iris[,"Petal.Width"])
```

That command uses a new strategy for identifying a column in a data set by its name rather than by a number. It produces the following output:

```
Pearson's product-moment correlation
data: iris[, "Sepal.Width"] and iris[, "Petal.Width"]
t = -4.7865, df = 148, p-value = 4.073e-06
alternative hypothesis: true correlation is not equal to 0
95 percent confidence interval:
-0.4972130 -0.2186966
sample estimates:
cor
-0.3661259
```

The interpretation of this result begins with stating the null hypothesis, that rho, the population correlation coefficient between sepal width and petal width, is zero. The alternative hypothesis is simply the logical opposite, and incorporates the possibility of a nonzero correlation that is either negative or positive. In fact, we get a hint right at the beginning of the output that this correlation is negative from the minus sign on the t-test. The observed value of t on 148 degrees of freedom is −4.79. Because the corresponding p-value, 4.073e-06, is decidedly less than the conventional alpha threshold of $p < .05$, we reject the null hypothesis. Remember that the scientific notation e-06 means that we should move the decimal point six spaces to the left to get the corresponding decimal number (0.00000473).

To become better informed about the uncertainty around the point estimate of our correlation, we can also look at the width of the confidence interval, which ranges from −0.497 up to −0.219. Although that is a fairly wide range, the confidence interval does not straddle 0, so we have a sense of certainty that the correlation is negative. In fact, the point estimate for the correlation reported by R is −0.36. If you check carefully you will find that $r = -0.36$ does not quite fall symmetrically between −0.497 and −0.219. This result is to be expected if you hearken back to our earlier discussion of the ceiling effect that is imposed on correlations because they can't go any higher than 1.0 or any lower than −1.0.

BAYESIAN TESTS ON THE CORRELATION COEFFICIENT

Experts have been working on versions of Bayesian tests that can directly examine the value of a Pearson's r correlation, but for now those procedures are not available in an R package. With that said, I used a little statistical trickery so that we can take advantage of the capabilities that do exist in the BayesFactor package to make our own Bayesian test of the correlation coefficient. The following code creates a custom function to do the job:

```
install.packages("BayesFactor")
library("BayesFactor")
bfCorTest <- function (x,y)  # Get r from BayesFactor
{
zx <- scale(x)                            # Standardize X
zy <- scale(y)                            # Standardize Y
zData <- data.frame(x=zx,rhoNot0=zy)      # Put in a data frame
bfOut <- generalTestBF(x ~ rhoNot0, data=zData)      # linear coefficient
mcmcOut <- posterior(bfOut,iterations=10000)         # posterior samples
print(summary(mcmcOut[,"rhoNot0"]))       # Show the HDI for r
return(bfOut)                             # Return Bayes factor object
}
```

The function() command in the code above lets R know that we want to create our own custom piece of reusable code. The x and y after the word "function" are names for the arguments that we pass to the function so it can do its work. Everything within the curly braces belongs to the function. The return() command at the end sends back a value to whoever called the function—normally we call it from the command line and it echoes the returned value to the console. In order to make a new function work, you have to either type all of this code at the command line or use R-Studio to "source" the function from the code window. Now whenever we want to test a correlation, Bayesian-style, we can call bfCorTest() and supply it with the names of any two variables. The two variables have to have the same number of observations. This custom function will report a point estimate and a 95% HDI for the correlation coefficient from the posterior population distribution. The function will also return a BayesFactor object, which could also be used for additional analyses. The Bayes factor—that is, the odds ratio—shows the odds in favor of the alternative hypothesis that the population correlation coefficient, rho, is not equal to 0. We can go back to our original example from the beginning of the chapter:

```
set.seed(12345)
wood <- rnorm(24)
heat <- rnorm(24)
bfCorTest(wood,heat)
```

Your output should look something like this:

```
Iterations = 1:10000, Thinning interval = 1 , Number of chains = 1
Sample size per chain = 10000
1. Empirical mean and standard deviation for each variable,
   plus standard error of the mean:
   Mean        SD    Naive SE   Time-series SE
-0.04535    0.18201   0.00182      0.00182
2. Quantiles for each variable:
   2.5%     25%      50%      75%     97.5%
-0.4105  -0.1622  -0.0420   0.0724   0.3140
```

Bayes factor analysis:
[1] rhoNot0 : 0.385294 ±0%
Against denominator: Intercept only
Bayes factor type: BFlinearModel, JZS

A thing of beauty! The point estimate for rho is −0.045, similar to what our exercise at the beginning of the chapter showed. The 95% HDI ranges from −0.411 up to 0.314, a truly huge range that squarely straddles 0. Finally, the Bayes factor of 0.385 means that the odds are actually in favor of the null hypothesis (though only weakly so). To make the 0.385 more intuitive, you can invert it like this, $1/0.385 = 2.6$, in other words 2.6:1 odds in favor of the null hypothesis. Using a rule-of-thumb odds cutoff value of 3:1, this is only anecdotal evidence in favor of the null hypothesis, but on the other hand, it does not in any way support the alternative hypothesis, so we can certainly conclude that the value of rho is near, if not exactly at, 0.

I cleverly labeled the Bayes factor "rhoNot0" to remind us that the alternative hypothesis that the Bayes factor is comparing to the null hypothesis is that the population value rho is different from zero. Remember that when seeking support for the alternative hypothesis we want this odds ratio to be as high as possible and certainly well in excess of 3:1. Next, let's rerun the analysis after forcing a correlation between our two variables. We can use the same method as before to create a new version of the heat variable:

```
newHeat <- wood/1.41 + heat/1.41
bfCorTest(newHeat, wood)
```

I won't reproduce the whole output this time, but you should be able to verify by running the code above that the point estimate for rho is 0.45, somewhat lower than the value produced by the conventional method. Correspondingly, the 95% HDI ranges from 0.0799 to 0.822, quite a wide span. Nonetheless, as this span does not encompass 0, we have a credible notion that rho is positive. This result is confirmed by the Bayes factor output, which shows odds of 7.81:1 in favor of the alternative hypothesis that rho is not equal to 0.

To conclude our consideration of the PPMC, let's analyze the iris data as we did above, but this time with the Bayesian correlation test. We can use our custom function again:

```
bfCorTest(iris[,"Sepal.Width"],iris[,"Petal.Width"])
```

I've taken the liberty of chopping out the parts of the output you have seen before. Here are the primary pieces of information we care about:

1. Empirical mean and standard deviation for each variable, plus standard error of the mean:

Mean	SD	Naive SE	Time-series SE
-0.3488995	0.0763712	0.0007637	0.0007997

2. Quantiles for each variable:

2.5%	25%	50%	75%	97.5%
-0.4994	-0.4007	-0.3481	-0.2977	-0.1994

Bayes factor analysis
[1] rhoNot0 : 3864.927 ±0%

This output shows a mean correlation in the posterior distribution of rho is −0.35, just slightly smaller in magnitude than the conventional result. The 95% HDI spans the range of −0.444 up to −0.199, just slightly wider than the conventional confidence interval. The Bayes factor of 3,865:1 in favor of the alternative hypothesis provides very strong evidence that the population correlation, rho, between Sepal.Width and Petal.Width, is not 0. Taking all of this evidence together, we can say with some credibility that the population correlation is a negative value lying somewhere in the range of −0.444 up to −0.199, and probably close to a central value of −0.35.

CATEGORICAL ASSOCIATIONS

The earlier material in this chapter focused on the association between two vectors of metric data, using the PPMC, which we usually just call the "correlation" and which we abbreviate as r. Not all data are metric, however, and another common analysis situation is where we have two categorical factors and we want to see if they are related to one another. Here we will return to the contingency tables that we explored in Chapter 3. As before, let's begin with a consideration of some random data where we would expect to find no association. To begin this exploration, I am bringing back the toast example from Chapter 3. My apologies, as you probably thought we were finally done with toast! This 2 × 2 contingency table is proportionally identical to the one in Chapter 3, except that I have included 10 times as many observations in each cell, as shown in Table 7.1.

As a reminder, based on this table, we know that there is an equal number of toast-down and toast-up events overall (50:50 in the marginal totals shown in the bottom row). We also know that only 30% of toast drops have jelly while 70% have butter (30:70 in the marginal totals shown in the right-hand column). The question we can ask with these data is whether the topping (butter

TABLE 7.1. Toasty Contingency Table with 100 Events			
	Down	Up	Row totals
Jelly	20	10	30
Butter	30	40	70
Column totals	50	50	100

vs. jelly) has anything to do with how the toast lands (down or up). In other words, is there an association between these two categorical variables, topping type and landing result? The logical opposite would be that they are **independent,** that is, there is no connection between topping type and landing result.

When we examined this question in Chapter 3, we concluded that there certainly must be an association because jelly has twice as many down-events as up-events, whereas butter goes the other way round (only three up-events per four down-events). What we did not consider in Chapter 3 is the idea that this particular contingency table happens to be just a single sample drawn from a population. What would happen if we repeated this experiment many times and examined the level of association in each one? Before we can go there, we need to have a sense of what the table would look like if there was absolutely no association between topping type and landing result. Consider the somewhat modified contingency table shown in Table 7.2.

There are a few important things to observe about this modified table. First, all of the marginal totals have stayed the same. I think that is interesting, as it suggests that the marginal totals are not the place where the action is happening with respect to the association between the two categories. Second, within the core of the table, the ratios between pairs of cells are now identical. For instance, for jelly toast, there are 15 up and 15 down and this ratio perfectly matches the one for butter—35 up and 35 down, in other words a 1:1 ratio in both cases. Likewise, if we compare in the other direction, for down events, the ratio of jelly to butter is 15:35, and that ratio is identical to the one for up events, 15:35. So intuitively we can see according to Table 7.2 that there is no association between topping type and landing result. This modified table represents the null hypothesis of indepedence, and is often referred to by statisticians as the *expected* values. You might be wondering how I came up with that modified table: There is a simple mathematical method for deriving the expected values from the marginal totals. Multiply the row total for a cell in a given row by the column total for that same cell, and divide by the grand total. For example, for jelly-down events you use (30 * 50)/100 to calculate the expected value for the upper-left cell in the expected values table.

There is one important subtlety at work here that I need to point out before we can proceed. Remember that the expected values come directly from the marginal totals. We start with a set of observations, each of which fits into

TABLE 7.2. "Null Hypothesis" Contingency Table—No Association			
	Down	**Up**	**Row totals**
Jelly	15	15	30
Butter	35	35	70
Column totals	50	50	100

one of the four cells of the contingency tables. But once we have calculated the marginal totals from raw data, almost everything about the original contingency table is now fixed. In fact, for a 2 × 2 table like this one, once you have calculated the expected frequencies from the marginal totals, only one cell in the original table is free to vary—one degree of freedom! This may seem weird, but you can reason it through yourself. Looking back at Table 7.1, let's hypothetically change the jelly-down cell in the upper left from 20 to 25. Then after we do that, the neighboring cell just to the right (jelly-up) must now shift down to 5 in order to hold the marginal row total at 30. Similarly, the neighboring cell just below (butter-down) has to shift down to 25 in order to hold the column total at 50. Finally, you can see that after all of those changes, the butter-up cell has to jump up to 45 to make everything else work out. Change any one cell, while holding the marginal totals the same, and all three of the other cells must change in response.

Another weird idea: if you play with these numbers a little bit yourself, you will see that there is a limit to how far the changeable cell can move. For example, if we push jelly-down up to 30, it forces jelly-up to 0 in order to maintain the row total. Since we are talking about counts of events in different categories, we cannot have fewer than 0! A negative event count? It makes no sense. Similarly, you cannot have any fewer than 0 jelly-down events, in part because the butter-down and jelly-up events are then maxed out in order to maintain the marginal row and column totals. So, keeping in mind that the expected value of jelly-down is 15, we can conclude that the minimum jelly-down is 0 and the maximum jelly-down is 30, given that we must maintain our exact marginal totals that were used to create the expected value table. Just to make it clear, here are the minimum (Table 7.3) and maximum (Table 7.4) tables.

I hope it is clear to you that either of these extreme scenarios represents a strong association between topping type and landing result. In Table 7.3, having jelly pretty much dictates that your dropped toast will land topping side up, whereas with butter you are more than twice as likely to land with the butter down. In Table 7.4 it is just the opposite. So if you imagine this situation in sampling terms, because there is only one *degree of freedom* in the table, we can sample values between 0 and 30 for just one cell, with an expected value of 15. Anything near 15 suggests that there is no association, whereas anything closer to either extreme would give stronger and stronger evidence of an association.

TABLE 7.3. Minimum Possible Value of Jelly-Down			
	Down	Up	Row totals
Jelly	0	30	30
Butter	50	20	70
Column totals	50	50	100

TABLE 7.4. Maximum Possible Value of Jelly-Down			
	Down	Up	Row totals
Jelly	30	0	30
Butter	20	50	70
Column totals	50	50	100

EXPLORING THE CHI-SQUARE DISTRIBUTION WITH A SIMULATION

So let's sample some values and see what we get. Before we start, we need a measure of how far the sampled contingency table varies from the expected values table. We could just subtract the expected table from the actual table. For example, in Table 7.4, the jelly-down of 30 minus the expected value of 15 gives a difference of 15. If we did this across all of the cells, though, we would get a mix of negative and positive numbers that would just cancel each other out. So instead, let's calculate the square of the difference between the actual value and the expected value for each cell. Then, so as to put each cell on a similar footing, let's divide the squared difference by the expected value. So for each cell we have ((actual - expected)^2)/expected. Sum the results that we get for each cell and we have a nice measure of how far any sampled contingency table varies from the expected values table. Statisticians call this quantity "chi-square" (you can also say, "chi-squared" if you like). Chi is the Greek letter that looks like an x. Karl Pearson, the originator of the Pearson product–moment correlation (PPMC), also had a substantial influence on the development of the chi-square test (Plackett, 1983), which we will use in the material below.

Creating a 2×2 contingency table in R would be easier with a simple R function. See if you can figure out how this function works:

```
# The user supplies the count for the upper-left
make2x2table <- function(ul)
{
ll   <- 50 - ul      # Calculate the lower-left cell
ur   <- 30 - ul      # Calculate the upper-right cell
lr   <- 50 - ur      # Calculate the lower-right cell
# Put all of the cells into a 2 x 2 matrix
matrix(c(ul,ur,ll,lr), nrow=2, ncol=2, byrow=TRUE)
}
```

This is pretty yucky code, as it hardwires the original marginal totals from Table 7.1 into the calculation of the other three cells, but I wanted to keep the list of arguments simple for this second example of creating a new function. Hardwiring constants like 30 or 50 into a function keeps it simple but limits its use to this one situation. The "ul" (which stands for "upper left") in parentheses in the first line lets us hand over one numeric value for the function to use. You will see ul show up three additional times in the other lines of code within the function. Let's test our new function to make sure we can reproduce the expected, minimum, and maximum tables we displayed above:

```
make2x2table(15)     # Should be like Table 7.2
     [,1]  [,2]
[1,]  15    15
[2,]  35    35


make2x2table(0)      # Should be like Table 7.3
     [,1]  [,2]
[1,]   0    30
[2,]  50    20


make2x2table(30)     # Should be like Table 7.4
     [,1]  [,2]
[1,]  30     0
[2,]  20    50
```

Next, let's also make a function that computes the chi-square value by comparing an actual and an expected values matrix.

```
calcChiSquared <- function(actual, expected)     # Calculate chi-square
{
diffs <- actual - expected              # Take the raw difference for each cell
diffsSq <- diffs ^ 2                    # Square each cell
diffsSqNorm <- diffsSq / expected       # Normalize with expected cells

sum(diffsSqNorm) # Return the sum of the cells
}
```

This function takes as its argument an actual and an expected values matrix. These matrices must have the same dimensions. In the case of our toast example, we will stick with a 2 × 2 matrix. R has intuitive matrix operations, so subtracting the contents of one matrix from another requires just a single step. We square the results and then normalize them by dividing by the expected value in each cell. Armed with these two functions, we are now in a position to do some sampling, but let's first retest our minimum, expected, and maximum frequencies for the upper-left cell.

```
# This makes a matrix that is just like Table 7.2
# This table represents the null hypothesis of independence
expectedValues <- matrix(c(15,15,35,35), nrow=2, ncol=2, byrow=TRUE)
calcChiSquared(make2x2table(15),expectedValues)     # Yields 0
calcChiSquared(make2x2table(0),expectedValues)      # Yields 42.86
calcChiSquared(make2x2table(30),expectedValues)     # Also yields 42.86
```

In the first command above, I created a 2 × 2 matrix containing the expected values (see Table 7.2) from our toast example. In the second command, I supplied the value of 15 to be used as the jelly-down value. As we know from the example, this is the expected value for that cell under the null hypothesis of no association, and supplying that value to the make2x2table() function dictates the other three values in the contingency table. Because the resulting table is identical to the expected values table, the resulting chi-square value is naturally 0. You should be able to see from the function definition for calcChiSquared() that the very first step will create a difference matrix where each cell is 0. The final two commands above show the chi-square value for the minimum and maximum legal values that we can put in the jelly-down cell. Each calculation comes out to 42.86. I think it is very cool that the chi-square value for either of these extreme differences is identical. Can you explain why they are identical?

Next, let's do some random sampling to simulate a range of different jelly-down counts that we might obtain. As we have done for other simulation exercises, we will begin with an assumption of the null hypothesis: that the jelly-down count is, over the long run, the "no association" value of 15. I will use the binomial distribution, as it provides discrete values that we can use as counts and we can easily center it at 15. Note that this will only give a rough approximation to the actual distribution that undergirds chi-square, but it will suffice for the purposes of demonstration. In these first two commands, I calculate a mean of the binomial distribution to double-check that it centers on 15 and I also request a histogram over 1,000 randomly generated values so that I can see the empirical distribution (see Figure 7.8). The second parameter in rbinom() is 30 because that is the number of trials (e.g., coin tosses or toast drops) that we need to do to get a mean of 15 (e.g., 15 heads-up, or 15 jelly-up).

```
set.seed(12)
mean(rbinom(1000,30,prob=0.5))
hist(rbinom(1000,30,prob=0.5))
```

The mean command above reports a mean of 14.9 for this distribution of 1,000 randomly generated values, which is close enough to 15 for the purposes of our simulation. Remember that we are exploring a simulation of the null hypothesis that the topping type and landing result are not associated. If you go back to our *expected values* table you will remember that the number of jelly-down events (the cell in the upper-left corner) has to be 15. So in Figure 7.8 we have created a plausible distribution of jelly-down events that centers on 15. Now we can replicate our process of calculating chi-square values based on this distribution of jelly-down events:

```
chiDist <- replicate(100000, calcChiSquared(make2x2table(rbinom(n=1,
    size=30,prob=0.5)),expectedValues))
```

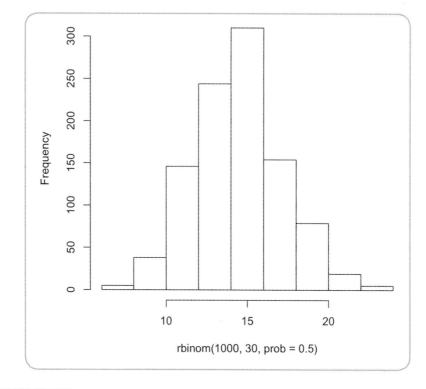

FIGURE 7.8. Histogram of 1,000 random data points from the binomial distribution with 30 events per trial.

In this command we have used replicate() to run a process 100,000 times. Each time we ask rbinom() to produce one random value in the binomial distribution depicted in Figure 7.8. That value, which we are treating as the count of jelly-down events, is piped into the make2x2table() function as the upper-left cell. Because there is only one degree of freedom, specifying the upper-left cell creates a complete 2×2 matrix. We then use that matrix together with a table of expected values to calculate the chi-square value. So the variable chiDist now contains an empirical distribution of chi-square values that we can examine using a histogram (see Figure 7.9).

```
hist(chiDist)
```

Next, using the standard logic of the null hypothesis test, we can divide the distribution into a main area containing 95% of cases and an "unusual" tail containing 5% of cases. Notice that this is an asymmetric distribution with no values less than 0. For this reason, we do not perform a two-tailed test the way we would with a symmetric distribution such as the *t* distribution. This is more like the *F*-test, which only has a positive tail. The quantile command reveals this cutoff point for null hypothesis testing:

```
quantile(chiDist, probs=c(0.95))
    95%
4.761905
```

Thus, using the logic of the null hypothesis test, any chi–square value above 4.76 would be considered quite unusual if the null hypothesis were "true." Keep in mind that this is just an informal simulation to demonstrate the concepts graphically: the actual critical value of chi-square on one degree of freedom is a little different than what the quantile() command shows. Now let's fiddle around a little bit with example contingency tables to explore how far off of the expected values our jelly-down events would have to be in order to qualify:

```
calcChiSquared(make2x2table(20),expectedValues)
[1] 4.761905
```

```
calcChiSquared(make2x2table(10),expectedValues)
[1] 4.761905
```

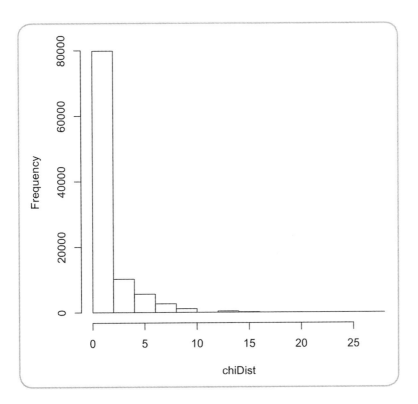

FIGURE 7.9. Histogram of 100,000 trials of calculating chi-squared, under the assumption that the null hypothesis is true.

Look back a few pages to make sure you remember how make2x2table(20) creates a 2 × 2 contingency table with a count of 20 in the upper-left corner. Similarly, make2x2table(10) produces a contingency table with 10 in the upper-left. We get the same chi-square value of 4.76 from either table. Putting it all together, if we provide jelly-down events that are five events or more higher or lower than the expected value of 15, the resulting chi-square would be right on that threshold of the tail of the empirical distribution of chi-square that appears in Figure 7.9. Now let's use the "official" significance test on chi-square. Here is the R code that performs the standard null hypothesis test, known as chisq.test():

```
chisq.test(make2x2table(20), correct=FALSE)
```

That command yields the following output:

```
Pearson's Chi-squared test
data: make2x2table(20)
X-squared = 4.7619, df = 1, p-value = 0.0291
```

Using our custom make2x2table() function, we have created a 2 × 2 contingency table with 20 jelly-down events (upper-left corner). We use that table to call chisq.test() which calculated chi-square from the table and obtains the associated p-value. The output shows df = 1, one degree of freedom, for the reasons I mentioned above: in a 2 × 2 table, once the marginal values are fixed only one cell is free to vary. More generally, for tables larger than 2 × 2, the number of degrees of freedom is (rows-1)*(cols-1). For example, a contingency table with three rows and four columns would have df = (3-1)*(4-1) = 6. The parameter "correct=FALSE" suppresses the so-called Yates correction, which would have made the chi-square test more conservative for small samples. Only use correct=TRUE if any one of the cells in the contingency table has fewer than five observations. The observed chi-square calculated from this table is 4.76 with an associated p-value of 0.029. Because that p-value is smaller than the typical alpha threshold of $p < .05$, we reject the null hypothesis. As an exercise, you should repeat the chisq.test() procedure on a matrix with 10 jelly-down events.

THE CHI-SQUARE TEST WITH REAL DATA

Let's close this section with an example that uses real data. I'm going to use a new function here, known as ftable(), that flattens contingency data into a table that we can use as input to chi-square. I am applying ftable() to the built-in "Titanic" data set, which contains a list of passenger categories and how many of each type of passenger survived the wreck. You should not need to type data(Titanic) at the command line, as the data should already be loaded, but it would be worthwhile to review the raw data by just typing Titanic at the command line. The following call to ftable() extracts a split of survivors and nonsurvivors by gender, which we then test for independence from survival status with the chi-square null hypothesis test:

```
badBoatMF <- ftable(Titanic, row.vars=2, col.vars="Survived")
badBoatMF
chisq.test(badBoatMF, correct=FALSE)
```

The second line of code, which reports badBoatMF to the console, gives the following output:

Survived	No	Yes
Sex		
Male	1364	367
Female	126	344

Clearly, if you had taken that fateful boat ride, it was not a good thing to be a guy. The output of the chi-square test strongly confirms that expectation:

```
    Pearson's Chi-squared test
data: badBoatMF
X-squared = 456.87, df = 1, p-value < 2.2e-16
```

The reported value of chi-square, 456.9 on one degree of freedom, has an extremely low corresponding p-value, well below the standard alpha level of $p < .05$. Thus we *reject the null hypothesis of independence* between gender and survival status. These two factors are not independent, and by inspecting the 2 × 2 contingency table we can see that the proportion of survivors among males was considerably lower than the proportion of survivors among females.

THE BAYESIAN APPROACH TO THE CHI-SQUARE TEST

The BayesFactor package contains a Bayesian test of contingency tables that produces a Bayes factor and that can optionally produce posterior distributions for the frequencies (or proportions) in the cells of the contingency table. The command is called contingencyTableBF().

The interface for contingencyTableBF() is similar to that for chisq.test(), but we need to make one important decision before we are ready to use it. You may recall some previous discussion of prior probabilities in our earlier discussions of Bayesian techniques. With contingency tables, the strategy used to collect the data affects the choice of priors. In the simplest case, where we collect data with no notion of how many cases we will end up with and no special plans for the marginal proportions, we use the parameter "sampleType=poisson" in our call to contingencyTableBF(). Jamil et al. (2016) commented that this choice is particularly appropriate for observational studies where there is no specific target for the number of observations or the proportions assigned to different categories. Our dropped toast study seems to qualify in that regard.

For the second option, "sampleType=jointMulti," the total number of observations is assumed to be fixed, but there is no specific target for proportions in categories. For example, we might conduct research on 100 representatives in a legislative body, where we would tally yes/no votes in the columns and the rows could represent whether or not the representative received a donation from a particular company. The total number of observations is fixed, because there are exactly 100 representatives, but the row and column proportions can vary freely.

The third option, "sampleType=indepMulti," assumes that the marginal proportions for either the rows or the columns are fixed. This situation might arise in an experimental study where we planned to assign half of the cases to one condition and half to the other condition. Finally, "sampleType=hypergeom," makes both row and column marginal proportions fixed. As you will recall from our discussion of degrees of freedom, with fixed marginal proportions we leave only one cell to vary (in a 2×2 design). Study designs with fixed row and column proportions are rare. The only possibility I could think of would be a situation where two metric variables were recoded with median splits. This would fix both the row and column margins at 50:50.

As mentioned above, our dropped toast study seems to best fit the Poisson priors, as we did not set out to have a fixed number of observations or a fixed proportion of toppings. Whether toast lands topping up or down is presumably random, as we have previously discussed. With all of that settled, let's apply the Bayesian contingency test to the original toast-drop data in Table 7.1 (for your convenience, I also repeat those same data here in Table 7.5). The appropriate call to contingencyTableBF() is:

```
ctBFout <- contingencyTableBF(make2x2table(20), sampleType="poisson",
    posterior=FALSE)
ctBFout
```

TABLE 7.5. Toasty Contingency Table with 100 Events			
	Down	**Up**	**Row totals**
Jelly	20	10	30
Butter	30	40	70
Column totals	50	50	100
Note. This table is the same as Table 7.1.			

The second command above echoes the contents of ctBFout to the console, displaying the Bayes factor as we have seen in previous uses of the BayesFactor package:

```
Bayes factor analysis
--------------
[1] Non-indep. (a=1) : 4.62125 ±0%
Against denominator:
Null, independence, a = 1
---
Bayes factor type: BFcontingencyTable, poisson
```

The Bayes factor of 4.62:1 is in favor of the alternative hypothesis that the two factors are not independent from one another (in other words, that the two factors are associated). Because the reported Bayes factor is in excess of 3:1, we can treat it as positive evidence in favor of nonindependence. Therefore, in this research situation, the Bayes factor and the null hypothesis concur with each other.

Are you up for a brain-stretching exercise? Because we can also get a 95% highest density interval from this test, but it will take a little bit of mental gymnastics and a little arithmetic to get there. First of all, we need to rerun the call to contingencyTableBF() to sample from the posterior distributions. That requires just a slight modification on the earlier command:

```
ctMCMCout <- contingencyTableBF(make2x2table(20),
    sampleType="poisson", posterior=TRUE, iterations=10000)
summary(ctMCMCout)
```

All we did there was to add two additional parameters to the call: posterior=TRUE and iterations=10000. The former causes the BayesFactor procedure to sample from the posterior distributions and the latter asks for 10,000 samples. As before, the number 10,000 is an arbitrary choice. If you have a fast computer you could ask for 100,000 or more and you will typically get somewhat more precision in your results. The resulting object, ctMCMCout, contains the results of the 10,000 samples in the form of means and HDIs for each of the cell counts. Here's slightly abbreviated output from the summary(ctMCMCout) command:

1. Empirical mean and standard deviation for each variable, plus standard error of the mean:

	Mean	SD	Naive SE	Time-series SE
lambda[1,1]	20.14	4.402	0.04402	0.04311
lambda[2,1]	29.85	5.310	0.05310	0.04948
lambda[1,2]	10.57	3.190	0.03190	0.03190
lambda[2,2]	39.44	6.202	0.06202	0.06202

2. Quantiles for each variable:

	2.5%	25%	50%	75%	97.5%
lambda[1,1]	12.47	17.03	19.83	22.84	29.73
lambda[2,1]	20.54	26.10	29.54	33.16	41.02
lambda[1,2]	5.271	8.30	10.24	12.51	17.78
lambda[2,2]	28.52	35.02	39.03	43.43	52.50

If you squint at the means in the first section, you will see that they closely match the contents of the cells in the original data. Go back to Table 7.5 to check on this. For example, Butter:Down, which is shown as lambda[2,1] (i.e., the mean population count for row two, column one) has a mean of 29.85 in the results above, whereas in the original data the value was exactly 30. So we don't really learn anything that we don't already know by looking at the means. Much more interesting (as usual) are the quantiles for each variable, including the boundaries of the 95% highest density interval (HDI) in the first and last columns. For example, even though the Butter:Down cell has both mean and median right near 30, the 95% HDI shows that the Butter:Down cell could be as low as 20.54 and as high as 41.02. Of course, those counts and any counts more extreme than those are fairly rare. We could get a more precise view of this by plotting a histogram of all the MCMC samples.

Before we go there, however, let me introduce one new idea to make the results more interesting. If you look back at Table 7.5, one way to reframe our research question about the independence or nonindependence of the two factors is to say that we want to know whether the *proportion of jelly to butter is different across the two columns*. In other words, if toast falls with the topping down, is the likelihood of it being buttered toast different than if the toast falls with the topping up? Now it should be obvious to you from Table 7.5 that, *in the one time we conducted this experiment, that is, only in this sample*, the proportions are apparently different. For down-facing toast the ratio is two jelly to every three butter (0.667), whereas for up-facing toast the ratio is one jelly to four butter (0.25). But we know from our inferential thinking that because of sampling error, this one sample does not precisely match the population. The virtue of the Bayesian results is that we have a range of possible cell counts for each cell to represent our uncertainty about the population. But instead of looking at the HDIs for individual cell counts, let's use the posterior distribution to create *ratios of cell counts*. First, we will do the left column of Table 7.5, in other words the jelly:butter ratio for down-facing toast:

```
downProp <- ctMCMCout[,"lambda[1,1]"] / ctMCMCout[,"lambda[2,1]"]
hist(downProp)
```

In the first command, we create a new list of 10,000 posterior results, each one calculated as a ratio of the Jelly:Down cell count to the Butter:Down cell count for one element in the posterior distribution. I have placed the result in a new vector called downProp, because we are going to use it later. The

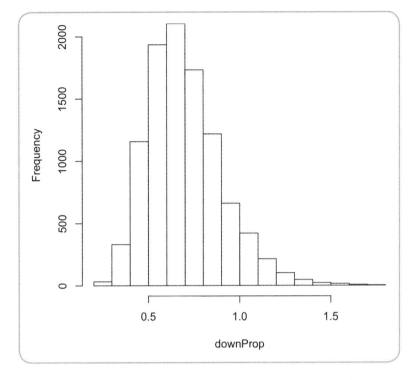

FIGURE 7.10. Posterior distribution of ratios of the Jelly:Down cell count to the Butter:Down cell count.

histogram of downProp, shown in Figure 7.10, shows a nice, fairly symmetric distribution with a mean that seems quite near to 0.667. You can verify this for yourself with the mean(downProp) command. Look at the range in that histogram. There are just a few posterior samples where the proportion of jelly to butter for down-facing toast was near 0 (indicating a very low count for jelly relative to butter). On the other hand, there are quite a few samples where the ratio is in excess of 1. For those samples, the count for the jelly cell was at least a little higher than the count for the butter cell. There's not much happening out at 1.5—that region represents a full reversal of the original ratio (3:2 instead of 2:3).

I'll venture you are getting the hang of this now: we can repeat that whole process for the left column of Table 7.5. Before you look at the resulting histogram, can you guess where it will be centered?

```
upProp <- ctMCMCout[,"lambda[1,2]"] / ctMCMCout[,"lambda[2,2]"]
hist(upProp)
```

The histogram in Figure 7.11 is centered on about 0.25, as we would expect from looking at the original data in the up-facing toast column of Table 7.5. Everything else in the histogram has shifted down a bit as well. We still

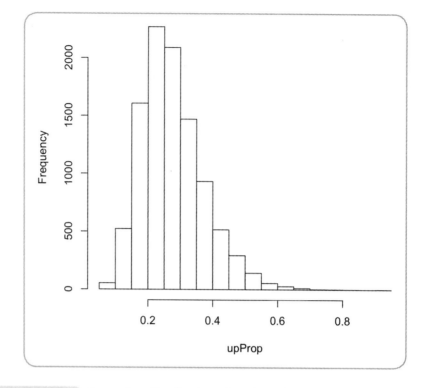

FIGURE 7.11. Posterior distribution of ratios of the Jelly:Up cell count to the Butter:Up cell count.

have 0 as the lower bound. This represents the smallest possible count for jelly events in the up-facing toast column. On the other hand, we have no posterior samples at all where the jelly count exceeded the butter count, as the histogram maxes out at about 0.8. By the way, you should run mean(upProp) just to verify where the exact center of the histogram lies.

And now for the fun part! Remember that what we really wanted to do was to compare the distribution of proportions for the left column versus the right column. In other words, as we switch our consideration from down-facing toast to up-facing toast, how much does the proportion of jelly to butter change? We can set this up with a very simple bit of arithmetic:

```
# The difference in proportions between columns
diffProp <- downProp – upProp
hist(diffProp)
abline(v=quantile(diffProp,c(0.025)), col="black")    # Lower bound of 95% HDI
abline(v=quantile(diffProp,c(0.975)), col="black")    # Upper bound of 95% HDI
```

This is so excellent: Figure 7.12 contains a histogram of the posterior distribution of differences in proportions between the two columns. To put this idea into different words, this is how much the jelly:butter ratio decreases as we switch columns from down-facing toast (left column) to up-facing toast (right column). The center of this distribution is a difference in proportions of 0.42. I've used abline() to put in vertical lines marking off the 95% HDI. The low end of the HDI is just barely above 0, while the top end of the HDI is just below 1. This evidence accords with both the Bayes factor and the null hypothesis test on chi-square: there is credible evidence that in the population there is an association between topping type and toast landing. In the population, the proportion of jelly to butter shifts by about 0.42, although there is a small likelihood that the difference in proportions could be as little as about 0.03 or as much as about 0.90.

OK, we are ready to finish up now with an analysis of some real data, so let's repeat our analysis of gender and survival on the Titanic:

```
badBoatMF <- ftable(Titanic, row.vars=2, col.vars="Survived")
ctBFout <- contingencyTableBF(badBoatMF,sampleType="poisson",
    posterior=FALSE)
ctBFout
```

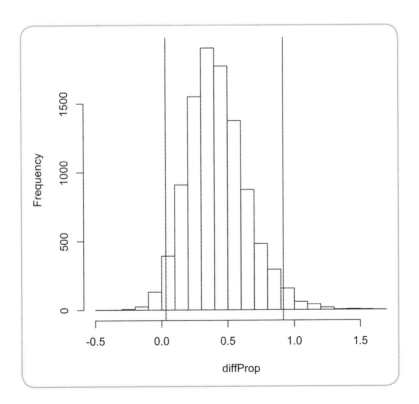

FIGURE 7.12. Histogram of the posterior distribution of differences in proportions between the two columns.

The Bayes factor arising from this analysis is 1.494287e+93, a truly massive number that is very strongly in favor of the alternative hypothesis of nonindependence. Not surprisingly, these results accord with the results of the null hypothesis test. I will leave the analysis of proportions to you as an exercise. Because I am an awesome guy, I will supply you with the R code to accomplish this analysis:

```
ctMCMCout <- contingencyTableBF(badBoatMF,sampleType="poisson",
    posterior=TRUE,iterations=10000)
summary(ctMCMCout)      # Review the posterior distributions of cell counts
# 1st row
maleProp <- ctMCMCout[,"lambda[1,1]"]/ctMCMCout[,"lambda[1,2]"]
# 2nd row
femaleProp <- ctMCMCout[,"lambda[2,1]"]/ctMCMCout[,"lambda[2,2]"]
diffProp <- maleProp - femaleProp   # The difference in proportions by gender
hist(diffProp)        # Histogram the distribution of differences in proportions
abline(v=quantile(diffProp,c(0.025)), col="black")    # Low end of 95% HDI
abline(v=quantile(diffProp,c(0.975)), col="black")    # High end of 95% HDI
mean(diffProp)      # Here's the center of the distribution
```

CONCLUSION

This chapter was all about associations between variables. We covered the two biggies: association between two metric variables and association between two categorical variables. The former is accomplished by means of a correlation coefficient and the latter by tests of independence on a contingency table. We used the PPMC as our method of analyzing the association between two metric variables. The null hypothesis and Bayesian tests on a PPMC coefficient help us to understand whether the coefficient is credibly different from zero. Likewise, with contingency tables the chi-square test of independence and a corresponding Bayes factor help us to understand whether any apparent differences in proportions are due to sampling error or are credible evidence of nonindependence.

Note that there are additional possibilities, both for methods of correlation and for ways of analyzing contingency tables. Try typing ?cor at the command line. Look at the different options for "method" that R presents in the help. Likewise, try typing ?fisher.test at the command line to examine the help file for Fisher's exact test of independence. There are also R packages that provide the capability to test an association between a metric variable and a categorical variable. Always more to learn!

I would be remiss if I did not mention somewhere along the line an important saying that researchers often repeat to each other: "Correlation is not causation." This idea is important because, armed with our new statistical tools, it is sometimes too easy to lose sight of what's really going on in the research situation. For example, in a physics classroom we could measure the number of hours per week that a student spends at the tutoring center and correlate that value with the student's grade on the final exam. If we found a significant, negative correlation, would that indicate that more tutoring causes lower grades? I hope not! Instead, there might be a hidden third factor that influences both the choice to spend time at the tutoring center and the grade on the final exam. Without additional information, we really don't know, so it is important when interpreting a correlation coefficient to be circumspect in drawing a conclusion about what the correlation signifies. When we get a significant (i.e., nonzero) correlation between two variables, we have evidence that the variables are connected, but we don't really know anything about why or how they are connected.

EXERCISES

1. A geologist took 48 samples of rock and measured the area of pores in the rock and the permeability of the rock (the extent to which fluid can pass through). The correlation came out at about $r = -.40$. Say briefly in your own words what that correlation value means. A statistician conducted a null hypothesis significance test on the data and then declared, "The test was significant at an alpha level of 0.05!" What else can you say about the geologist's data now that you know the results of the null hypothesis significance test?

2. The R data sets() package actually contains a data set called "rock" that has data something like that described in exercise 1. Run cor(rock) and make sense of the results. Why are there two copies of the correlation between "area" and "perm"? What is meant by the numbers on the diagonal?

3. Run cor.test() on the correlation between "area" and "perm" in the rock data set and interpret the results. Note that you will have to use the "$" accessor to get at each of the two variables (like this: rock$area). Make sure that you interpret both the confidence interval and the p-value that is generated by cor.test().

4. Create a copy of the bfCorTest() custom function presented in this chapter. Don't forget to "source" it (meaning that you have to run the code that defines the function one time to make R aware of it). Conduct a Bayesian analysis of the correlation between "area" and "perm" in the rock data set.

5. Bonus research question: cor.test() does not work on a whole data set of values—it cannot display significance tests for a whole correlation matrix. Can you find and install a package that will calculate a whole matrix of significance tests on a numeric data set? Once you find and install the package, run the appropriate procedure on the rock data and report the results.

6. Bonus statistical thinking question: Run set.seed(123456) to set the random number seed and then run cor.test(rnorm(100), rnorm(100)). Describe what this cor.test()

command accomplishes with these particular inputs. Describe and interpret the output from this analysis. What cautionary tale is implied by these results?

7. Let's say a department within a university admits 512 male students and 89 female students. At the same time, they reject 313 male students and 19 female students. Put those numbers in a 2 × 2 contingency table. Calculate marginal totals. Use the marginal totals to construct a new 2 × 2 table of expected values. Describe in words what you see in these two tables. You should be able to say some things about the admission rates, about the ratio of female to male applicants, and possibly something about whether males or females are more likely to be admitted.

8. Not unexpectedly, there is a data set in R that contains these data. The data set is called UCBAdmissions and you can access the department mentioned above like this: UCBAdmissions[, ,1]. Make sure you put two commas before the 1: this is a three dimensional contingency table that we are subsetting down to two dimensions. Run chisq.test() on this subset of the data set and make sense of the results.

9. Use contingencyTableBF() to conduct a Bayes factor analysis on the UCB admissions data. Report and interpret the Bayes factor.

10. Using the UCBA data, run contingencyTableBF() with posterior sampling. Use the results to calculate a 95% HDI of the difference in proportions between the columns.

## CHAPTER 8

Linear Multiple Regression

You are not going to believe this, but most of the analysis techniques that we have discussed so far fit into one big family known as the *general linear model*. ANOVA and Pearson correlation are very much in the family, and even the chi-square test could be considered a cousin. The general linear model builds from the idea that there is one outcome variable, sometimes called a *dependent variable* and that this dependent variable is modeled as a function of one or more (usually more) *independent variables*. Each independent variable may be metric, ordered, or categorical. Here's a simple example that demonstrates the idea:

GPA = (B1 * HardWork) + (B2 * BasicSmarts) + (B3 * Curiosity)

Stated in words, this equation suggests that a student's grade point average (GPA) is a linear function of hard work plus basic smarts plus curiosity. For each of the independent variables there is a coefficient (labeled here B1, B2, and B3), what you might think of as either the slope in a line drawing or an index of the importance of the given independent variable. A statistical procedure such as "least-squares" analysis is used to calculate the value of each of the B coefficients.

Once calculated, the B coefficients can be used for several things. First, one could use them to create an equation like the one above that could be used to predict an unknown GPA of a student based on measurements of her hard work, basic smarts, and curiosity. This would be considered a **forecasting** application. Forecasting can be valuable for many things: for example, if we were trying to develop an educational early warning system that would let us know in advance that a student was likely to obtain a low GPA. Then we could provide some assistance, such as time management training, that would help the student work smarter.

More commonly, and especially in research applications, we are interested in knowing the magnitude of the B coefficients so that we can draw some conclusions about the relative importance of each of the predictors. In research, this kind of analysis contributes to an understanding of theory-based questions by drawing our attention to promising precursors of the dependent variable (which may or may not be direct causal influences on the variable). In applications to management and other kinds of decision making, we can use the B coefficients to select promising metrics that can serve as indicators of important processes. The key point is, that in both research and practice, the outputs of the regression analysis may have intrinsic value for our work even if we never intend to create a forecasting equation.

By far the easiest metaphor for understanding regression lies in the process of discovering a best-fitting line between two metric variables. As we did for correlation, we can learn a lot by modeling some simple random variables before we get to the process of exploring real data. You may remember how we created correlated variables before: now we are going to extend those techniques to explore multivariate regression. We begin by creating three random, normally distributed variables:

```
set.seed(321)
hardwork <- rnorm(120)
basicsmarts <- rnorm(120)
curiosity <- rnorm(120)
```

As usual, we use set.seed() so that our results will be reproducible. Within the limits of rounding, you should find the same results throughout this example if you use the same seed to start your random number sequence. We are going to pretend that we have a sample of $n = 120$ observations from students on the variables described earlier in the chapter. Next, we will synthesize a dependent variable by adding together these components. To make the example more realistic, we will also include a random noise component. This reflects the reality in most modeling situations, particularly in social science, management, and education: we never achieve perfect prediction when measuring people. Here's the code:

```
randomnoise <- rnorm(120)
gpa <- hardwork/2 + basicsmarts/2 + curiosity/2 + randomnoise/2
sd(gpa)
```

The second line creates a linear combination of the three independent variables plus a random noise component. We divide each input variable by 2 because it is the square root of 4 and we have four inputs into creating our fake dependent variable. This ensures that the standard deviation of GPA will be approximately one: The output of the sd() command confirms that. Next, let's begin our exploration by focusing on just one predictor variable: hardwork.

A very important tool for this bivariate situation is the *scatterplot* or what some people call the "scattergram." You should try to develop your expertise at "reviewing" a scatterplot to look for at least three important characteristics:

1. **Linearity:** a relationship between two variables will appear linear if you can imagine drawing a nice straight line that roughly fits the shape of the cloud of points in the scatterplot;

2. **Bivariate normality:** points in the scatterplot will be somewhat more dense near the point in the plot where the mean of the *X*-axis variable and the mean of the *Y*-axis variable intersect; likewise, points at the extreme ends of the scatterplot should seem more sparse; it is possible to model nonnormal variables using linear multiple regression, but one must take extra care when evaluating results;

3. Influential **outliers:** modeling with linear multiple regression is quite sensitive to extra outliers, especially at the high or low ends of the bivariate distributions; one lonely point off in a corner of the plot can have undue influence on the outputs of the analysis.

Figure 8.1 arises from plot(hardwork, gpa). See if you can spot the linearity, bivariate normality, and absence of outliers in the scatterplot that appears in the figure.

I think you should be able to see quite readily that the relationship looks linear as it extends from the lower left to the upper right, that many of the points cluster around the point {0,0}, which you can easily verify is roughly the mean of hard work and the mean of GPA. Finally, it should also be obvious from the plot that there are no extreme outliers. Because we created both the *Y* variable (gpa) and the *X* variable (hardwork) by sampling randomly from a standard normal distribution, all of these conditions are to be expected. Any large random sample of standard normal data points will have these "well behaved" characteristics and, of course, we synthesized gpa so that it would be correlated with hardwork. Feel free to experiment with adding outliers (just change one value of gpa to 10) or with using different random distributions (e.g., uniform or Poisson) to create one or more of the variables. If you redo the scatterplot, you should be able to see the differences these changes make.

Next, based on my eagle eye and a little bit of insider knowledge, I am going to add my personal guess for a line that would fit this scatterplot, using abline(a = 0, b = 0.56). The abline() function places a line on the most recent graph you have requested from R. The "a" input to this function is the intercept—because both means are 0 we can assume that the *Y*-intercept is also close to 0 (note that in these graphs the *Y*-intercept lies right in the center of the figure). The "b" input is the slope and I have supplied my guess for the slope, which is 0.56. The result appears in Figure 8.2.

Next, look at the point that is at the bottom left of the diagram. There are actually two points right near each other {-2,-2} and you can think about either

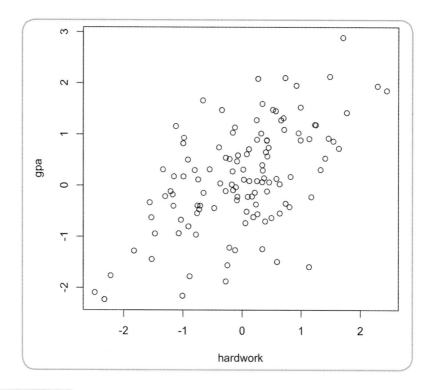

FIGURE 8.1. Scatterplot of two correlated random normal variables.

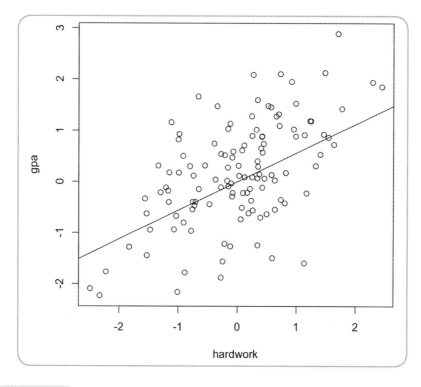

FIGURE 8.2. Scatterplot of two correlated random normal variables with an informed guess for the best-fitting line.

one. In your mind's eye, draw a vertical line from that point straight upward to the diagonal line that I drew on the diagram with abline(). This is the *error of prediction* for that particular point. If you are having trouble imagining the line that I am describing, try this command to add a little arrow to the plot:

```
arrows(min(hardwork),gpa[which.min(hardwork)],min(hardwork),
    min(hardwork)*0.56)
```

If you think about all the little vertical lines you could draw, you will notice that the supposedly "best-fitting" diagonal line is a little too high for some points, a little too low for others, and just about right for a few of the points. So, for every point on the plot, there is a vertical distance between the point and the best-fitting line. That distance is the **error of prediction** for that point. Let's look at all of those errors of prediction together and see what they look like. For each point in the data set, I am going to use a slope of 0.56 and an intercept of 0 to make a prediction of gpa (the Y value) from hardwork (the X value). For example, the point close to the line at the upper right of Figure 8.2 is {2.443,1.852}. To get the predicted value of gpa from this, I would first calculate gpa = 0.56*2.443 + 0. In this equation, the 2.443 is the X-value, the 0.56 is the slope, and the 0 is the Y-intercept. Run these numbers and the predicted value of gpa is 1.368. The difference between the observed value of gpa and the predicted value is the error of prediction, so $1.852 - 1.368 = 0.484$.

I can calculate errors of prediction for all of our made-up data points and our "guesstimated" best-fitting line using this method. The appropriate expression in R is simply: gpa-(hardwork*0.56). I am going to plot a histogram of all of those errors and get a sum of them as well (see Figure 8.3).

```
hist(gpa - (hardwork * 0.56))
sum(gpa - (hardwork * 0.56))
```

It is important that you get what this expression is doing: gpa - (hardwork * 0.56). The part in parentheses is the prediction equation. If every point fell perfectly on a line with 0.56 as the slope and 0 for the Y-intercept, then this expression would produce perfect predictions of gpa. We know that the prediction is imperfect and by subtracting (hardwork * 0.56) from the actual value of gpa, we learn the error of prediction for each hardwork and gpa pair. Notice in Figure 8.3 that the prediction errors are pretty much normally distributed and centered on 0. Intuitively, this is what we would expect for a best-fitting line: most of the prediction errors ought to be pretty small, although there will always be just a few errors that are fairly large. Some of the errors will be above the line and some will be below the line, but they should be pretty well balanced out. In fact, if my guesses about the slope and intercept had been perfect (which they weren't), the errors ought to sum up to 0. In fact, the sum of the errors came out to a little higher than 19, which isn't too bad given that we have 120 points in our data set.

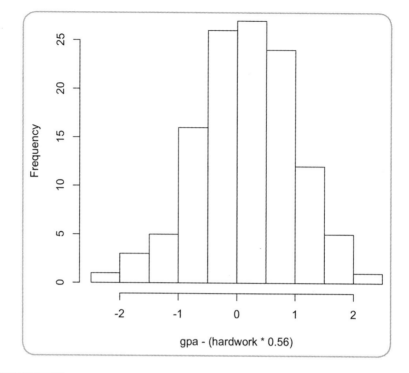

FIGURE 8.3. Histogram of prediction errors from points to the estimated line of best fit.

Of course, if we were trying to figure out how good our fitted line was by looking at the errors as a complete set, the fact that errors of prediction usually sum to 0 presents a little problem for us, analogous to when we were trying to calculate a measure of variability based on deviations from the mean. Intuitively, the smaller the errors are as a set, the better our prediction, but that is not helpful if all the positive errors just cancel out all the negative ones. Remember how we addressed that problem before? When we calculated variance (and its cousin, standard deviation), we squared each of the deviations to get rid of the sign. That's what we will do here as well. I have created a tiny function to do the job:

```
calcSQERR <- function(dv, iv, slope)
{
(dv - (iv*slope))^2
}
```

You might still be wondering about the details of how these functions work. The arguments listed in the function() definition—namely dv, iv, and slope—are passed in "by value" (i.e., copies of the arguments are made for the function to use). The output of the final evaluated expression in the function() is the value that the function returns. You can also use the return() function,

as we have done previously, to be very clear about the data object you want to
return. So this new function, calcSQERR(), calculates a vector of values that
are the squared errors of prediction based on predicting each observation of the
dv (dependent variable) from each observation of the iv (independent variable)
with the supplied value of the slope. This function is too dumb to include the
intercept if it is nonzero, so it will only work when we can assume that the
intercept is 0. This dumb function also assumes that the input vectors dv and
iv are exactly the same length. Let's try it using my guess for slope. I used the
calcSQERR() function with head() to review the first few values:

```
head( calcSQERR(gpa,hardwork,0.56) )     # head() shows the first few values
[1] 3.742599e+00 8.443497e-06 2.959611e+00 1.432950e+00
[5] 1.451349e+00 3.742827e+00

sum( calcSQERR(gpa,hardwork,0.56) )
[1] 86.42242

hist( calcSQERR(gpa,hardwork,0.56) )
```

The distribution that appears in Figure 8.4 should not be surprising: we
have gotten rid of all of the negative errors by squaring them. Still, most squared

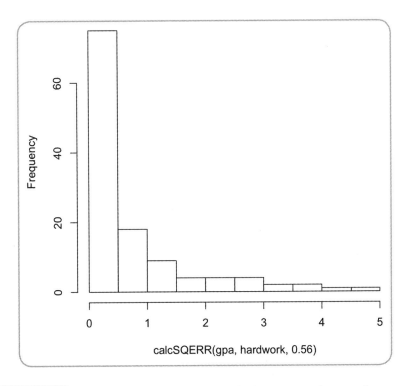

FIGURE 8.4. Histogram of squared prediction errors from points to the
estimated line of best fit.

errors are quite small, so they are near 0. You can also see that they are all positive (as a reminder, the e-06 stuff is scientific notation). I also calculated the
sum of squared errors of prediction, which is an important number for
capturing the total amount of error that results from our best-fitting line. In
fact, now that we have a function for calculating the squared errors of prediction, we can experiment with a range of different values of the slope to see if we
can find one that is better than my guess. Let's build a little wrapper function
for calcSQERR() to make this easier:

```
sumSQERR <- function(slope)
{
sum(calcSQERR(gpa, hardwork, slope))
}
sumSQERR(0.56)
[1] 86.42242
```

In the last line of code that says sumSQERR(0.56), I have tested the function and found that it produces the correct result for my current guess where
slope is 0.56. Like so many of my other functions in this book, this new function represents really bad programming practice because it embeds the variables
that we previously established, gpa and hardwork, right into the function. My
apologies to the good coders out there! On the other hand, by having just a
single input argument, slope, we greatly simplify the call to sapply() that we are
just about to do:

```
# A sequence of possible slopes
trySlopes <- seq(from=0, to=1, length.out=40)
# sapply() repeats calls to sumSQERR
sqerrList <- sapply(trySlopes, sumSQERR)
plot(trySlopes, sqerrList)        # Plot the results
```

In the first line of code we just create an evenly spaced sequence from
0 to 1, with 40 different values of slope to try out. After you run this line of
code, you can examine the contents of trySlopes to verify what it contains.
You should find 40 different values between 0 and 1: we use these as input to
the next two lines of code and you should think of them as lots of guesses as to
what the actual best-fitting slope might be. Hopefully, you can have an intuitive sense that a slope of 0 would represent no relationship between gpa and
hardwork, and that the sum of the squared errors with a slope of 0 ought to be
at a maximum. Additionally, a slope of exactly 1 might work perfectly if the
relationship between the two variables was also perfect, but we know that it is
not in this case. So the sweet spot—the value of slope that generates the smallest
sum of squared errors of prediction—should be somewhere between 0 and 1. If
my original guess was good, the slope of the line with the least-squared errors
should be right near 0.56.

In the second line of code above we use the cool and powerful sapply() function to repeatedly run our new custom wrapper sumSQERR(). By supplying trySlopes as the input variable for sapply(), we can get our custom function to run 40 times. Each of the 40 sum-of-squared-error values that is returned from sumSQERR will be put in a vector for us. We store that vector in a new variable, sqerrList. Try pronouncing that variable name, it sounds like squirrels. Finally, we create a plot that will show the different slope values we tried (X-axis) versus the resulting sum-of-squared-errors of prediction (Y-axis). What do you expect the shape of the resulting plot to be? Remember that we are expecting errors to be worst (the largest sum) with a slope of 0 and best (the smallest sum) somewhere near a slope of 0.56. The result appears in Figure 8.5.

Beautiful! Figure 8.5 looks like a parabola, which you might expect when working with a function where we squared the X values. It should be pretty obvious that the minimum value of the sum of squared errors of prediction occurs at a value of the slope that is just about 0.60. So my guess of 0.56 was pretty good, but off by a little bit.

What I have demonstrated here is a simple, mechanical method of using the **least-squares criterion** to find the best-fitting line. The best-fitting line in a bivariate relationship is the one with a value of slope and a value of

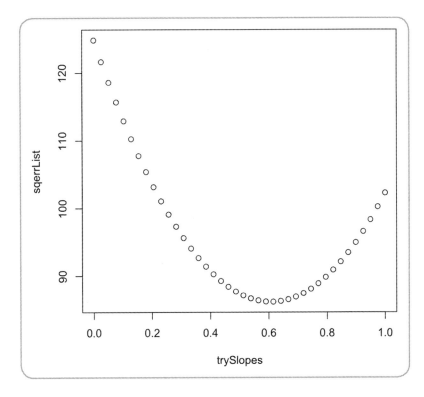

FIGURE 8.5. Minimizing the sum of squared errors of prediction by trying out different values of the slope of the best-fitting line.

intercept that, when used together on all of the data points, minimize the sum of the squared errors of prediction. Finding the slope and intercept that satisfies the least-squares criterion can be accomplished with a set of techniques that mathematicians call **matrix algebra.** Matrix algebra provides methods of multiplying big rectangles of numbers (and lots of other operations too). Given a particular data set and a specific prediction formula, there is one mathematical result produced by this matrix algebra procedure. The R command we use below employs this method to find the precise values of slope and intercept that optimize the least-squares criterion by minimizing the squared errors of prediction.

Now that we understand a little of what is happening "under the hood," let's use the lm() procedure in R to produce some "official" results. The lm() procedure uses the same model specification syntax as aov(); this model description syntax is pretty standard across all of R. In general, we put the dependent variable first, use the ~ character to separate it from the independent variables and then list out the independent variables, using + or * depending upon the kind of model we are building. Note that the first line in the code below constructs a data frame for us to offer to the lm() procedure. We then use the model syntax with ~ to let lm() know what we want to include:

```
# Put everything in a data frame first
educdata <- data.frame(gpa, hardwork, basicsmarts, curiosity)
regOut <- lm(gpa ~ hardwork, data=educdata)    # Predict gpa with hardwork
summary(regOut)                                # Show regression results
```

This code produces the following output:

```
Call:
lm(formula = gpa ~ hardwork, data = educdata)
Residuals:
```

Min	1Q	Median	3Q	Max
-2.43563	-0.47586	0.00028	0.48830	1.90546

Coefficients:

	Estimate	Std.	Error	t value	Pr(>|t|)
(Intercept)	0.15920	0.07663	2.078	0.0399	*
hardwork	0.60700	0.08207	7.396	2.23e-11	***

```
---
Signif. codes: 0 '***' 0.001 '**' 0.01 '*' 0.05 '.' 0.1 ' ' 1
Residual standard error: 0.8394 on 118 degrees of freedom
Multiple R-squared: 0.3167, Adjusted R-squared: 0.311
F-statistic: 54.7 on 1 and 118 DF, p-value: 2.227e-11
```

The first two lines repeat the function call used to generate this output. This can be very useful for making sure you know which output you are looking at and that R tested the model you wanted. Next, the summary of **residuals** gives an overview of the errors of prediction. Residuals and errors of prediction are exactly the same thing. The fact that the median is almost precisely 0 suggests that there is no skewness in the residuals, so the residuals are symmetrically distributed—as they should be.

The coefficients section begins to show us the key results we need to know. The first column is "Estimate" and that shows the intercept in the first line and the slope in the second line. Although we are working with samples of standard normal variables, it is clear that there is a little offset in the gpa data, because the intercept is at almost 0.16, just above the Y-axis.

Perhaps more importantly, the slope is estimated at about 0.61, slightly above my guess of 0.56, but definitely in the same ballpark. The standard errors around the estimates of slope and intercept show the estimated spread of the sampling distribution around these point estimates. The t-value shows the Student's t-test of the null hypothesis that each estimated coefficient is equal to zero. In both cases the value of t and the associated probability clearly indicates that we should reject the null hypotheses. You may note the asterisks that follow the probability values: as previously mentioned, these are a conventional way of marking different levels of alpha. One asterisk is used to indicate significance at $p < .05$, two for $p < .01$, and three for $p < .001$. Some statisticians have commented that including these conventional markers of the NHST reinforces the undesirable mechanical nature of the NHST (Meehl, 1978), because it encourages researchers to look at significance in isolation from other information. We may want to begin our interpretation by paying more attention to the effect size, which for regression is the R-squared value. The multiple R-squared for this analysis is just under 0.32, which means that hardwork accounted for about 32% of the variability in gpa. Not bad for a single predictor model!

Finally, the last three lines of the output show summary statistics for the performance of the whole model (albeit a rather simple bivariate model in this case). The standard error of the residuals is shown as 0.8394 on 118 degrees of freedom. Starting with 120 observations, one degree of freedom is lost for calculating the slope (the coefficient on hardwork) and one is lost for calculating the Y-intercept. The standard error of the residuals is calculated by starting with the sum-of-squared errors of prediction, dividing by the degrees of freedom, and taking the square root of the result. Finally, there is an F statistic that tests the null hypothesis that R-squared is equal to 0. Clearly, based on these results, we can reject that null hypothesis.

What would happen if we added more predictors to the model? If you add a second predictor, it is still pretty easy to visualize what is happening. Instead of a scatterplot of points on a two-dimensional surface, we would have a three-dimensional cloud of points. Instead of a best-fitting line, we would have a best-fitting plane. R has graphics packages that can show 3D plots, but we must save that exercise for another day. Once we get beyond two predictors, our ability to think of a geometric analog of the model breaks down, but in theory we can have any number of predictors in our model. If you remember the original setup of the data, we synthesized gpa to be a function of three predictor variables. Let's add the other predictors to the model and see what results we get:

```
regOut3 <- lm(gpa ~ hardwork + basicsmarts + curiosity, data=educdata)
summary(regOut3)
```

The formula uses gpa as the dependent variable—in regression, some researchers refer to this as the "criterion," while others call it the "outcome." All of the names after the ~ character are the independent variables—in multiple regression most researchers refer to them as "predictors." The results of that call to lm() produces the following output to the console:

```
Call:
lm(formula = gpa ~ hardwork + basicsmarts + curiosity, data = educdata)
Residuals:
     Min       1Q    Median       3Q       Max
 -1.02063  -0.37301  0.00361  0.31639   1.32679
Coefficients:
              Estimate   Std.    Error   t value   Pr(>|t|)
(Intercept)   0.08367   0.04575  1.829    0.07       .
hardwork      0.56935   0.05011  11.361   <2e-16    ***
basicsmarts   0.52791   0.04928  10.712   <2e-16    ***
curiosity     0.51119   0.04363  11.715   <2e-16    ***
---
Signif. codes: 0 '***' 0.001 '**' 0.01 '*' 0.05 '.' 0.1 ' ' 1
Residual standard error: 0.4978 on 116 degrees of freedom
Multiple R-squared: 0.7637,   Adjusted R-squared: 0.7576
F-statistic: 125 on 3 and 116 DF, p-value: < 2.2e-16
```

This is pretty exciting: the first time we have built a truly multivariate model! Using R's model specification syntax, we specified "gpa ~ hardwork + basicsmarts + curiosity" which reads in plain language as roughly this: Use a linear combination of the three independent variables—hardwork, basicsmarts, and curiosity—to predict the dependent variable gpa. The section of the output labeled "coefficients" is the first part to examine. In addition to the intercept, which we saw in the simple bivariate model as well, we have three coefficients, or what we will refer to as "B-weights." These B-weights are the slopes and together they describe the orientation of the best fitting hyperplane, that is, the linear geometric shape that minimizes the errors of prediction in our four-dimensional model space. Notice that the B-weights, 0.57, 0.53, and 0.51, are all quite similar, signifying that each of the predictor variables has roughly the same contribution to dependent variable. This is to be expected, because we synthesized the dependent variable, gpa, from an equal contribution of each of these predictors. The *t*-values and the probabilities associated with those *t*-values—the third and fourth columns of the coefficient output—test the null hypothesis that each of the coefficients is equal to zero. Obviously with such large *t*-values and small *p*-values, we can reject the null hypothesis in all three cases.

Making Sense of Adjusted *R*-Squared

If you thumb your way back through the book to "More Information about Degrees of Freedom" on p. 99, you will find a discussion of the meaning and uses of degrees of freedom. In a nutshell, in order to have an unbiased sample-based estimate, we have to take into account the degrees of freedom in the calculation of that estimate. For example, when we use $n-1$ in our calculation of sample variance, we make sure that, over the long run, these estimates of the population variance do not come out too low. Degrees of freedom helps account for the uncertainty in a sample, and makes sure that, over the long run, we don't capitalize on chance.

The same concept applies to the idea of adjusted *R*-squared. Remember that plain old *R*-squared simply represents the proportion of variance accounted for in the dependent variable: there are two variance estimates that make up that ratio. Plain old *R*-squared uses biased variance estimators, so over the long run the *R*-squared values estimated from samples are somewhat larger than the true population value.

When we calculate *adjusted R*-squared, however, we use unbiased estimators for those two variances. The total variance in the dependent variable is easy to figure: sum of squares divided by $n-1$ degrees of freedom in the denominator. But the proper degrees of freedom for the other variance is $n-p-1$, in other words, the sample size, minus the number of predictors, minus 1. You can think of that as a penalty for having more predictors in the model. Each additional predictor added to the model capitalizes on chance just a little bit more. Here's the formula for adjusted *R*-squared:

$$R_{adj}^2 = 1 - \left(\frac{SS_{res} / (n - p - 1)}{SS_{tot} / (n - 1)} \right)$$

This equation says that adjusted *R*-squared is 1 minus the ratio of two variances. Each variance is expressed as a sum of squares divided by the appropriate degrees of freedom. In the numerator we have the sum of squares for the residuals, while in the denominator we have the total sum of squares (for the dependent variable). You can see that the degrees of freedom for the residuals is $(n-p-1)$, that is, the sample size, minus the number of predictors, minus 1. The more predictors, the lower the degrees of freedom and the higher the estimate of the residual variance. By the way, the residual variance is sometimes also called the "error variance."

You might now be wondering when this adjusted *R*-squared matters and how it is used. The $n-p-1$ gives us our clue: whenever the sample size is small relative to the number of predictors, the downward adjustment will be substantial. For example, with a sample size of 20 and 9 predictors, the degrees of freedom for the residual variance would plummet down to $df = 10$. In turn, this will make the adjusted *R*-squared notably smaller than plain old *R*-squared. In this kind of situation, it is

(continued)

really important to report the adjusted *R*-squared so that readers do not interpret your results too optimistically.

In contrast, imagine if we had a sample size of 2,000 with 9 predictors. In that case, our degrees of freedom on the residual variance would be *df* = 1,990. This very slight reduction in *df* (due to the 9 predictors) would have almost no effect on the calculation of residual variance. The adjusted *R*-squared would be virtually identical to the plain old *R*-squared. In this kind of situation, it doesn't matter which one you report, because the plain old *R*-squared will not mislead people. You should always feel free to report the adjusted *R*-squared, as it paints a realistic picture of the performance of your prediction model regardless of sample size or the number of predictors.

In the model summary, the multiple *R*-squared value of 0.76 shows that about three-quarters of the variability in gpa was accounted for by the three predictor variables working together. *R*-squared can be interpreted as the proportion of variance in the dependent variable that is accounted for by a model. Again, the value of 0.76 is just what we would expect: You will recall that when we synthesized gpa it contained equal contributions from the three predictors plus a random noise component—all of which had equivalent standard deviations of about one. Note that there is an *F*-test (a null hypothesis significance test) of whether the multiple *R*-squared value is significantly different from 0. In this case we reject the *null hypothesis* because the probability associated with the *F*-test is well underneath the conventional alpha thresholds of $p < .05$, $p < .01$, and $p < .001$. See "Making Sense of Adjusted *R*-Squared" for a discussion of the adjusted *R*-squared.

Whatever happened to that random noise component that we injected into gpa, anyway? It mainly ended up in the residuals—the errors of prediction that are calculated for each case in the data set. You can access a complete list of the residuals using the residuals() command on the output of the regression analysis:

```
summary(residuals(regOut3))
    Min.    1st Qu.    Median      Mean    3rd Qu.       Max.
-1.021000  -0.373000  0.003611  0.000000  0.316400   1.327000
```

The mean of the residuals from the least-squares fitting process is always 0. That follows naturally from the definition of the best-fitting line (or plane or hyperplane), because the best-fitting line passes through the cloud of points in such a way as to have a balance of negative and positive errors of prediction. The closeness of the median to the mean shows that distribution of the residuals is not skewed. You should also run a histogram on the residuals to see the actual shape of the distribution. This is a very valuable exercise, because when a set of residuals is notably nonnormal, it indicates that the underlying relationships between the dependent variable and the independent variables are not linear. Multiple regression is a poor choice for modeling nonlinear relationships, at least without transformations applied to the variables, so make sure that you explore further if you have variables that are related by curved relationships or other nonlinear connections.

Just to close the loop on the thinking we are doing about regression diagnostics, Figure 8.6 has a scatterplot of the residuals plotted against the original random noise variable that we used to help synthesize gpa. The command plot (randomnoise, regOut3$residuals) will generate that graph for you. In a normal research situation, we don't have access to the original noise variable, but because of the fake variables we are using here, we have the opportunity to learn from it.

You will notice that there is a nearly perfect correspondence between the original random noise variable and the residuals from the regression analysis. You may wonder, however, why this plot is not a perfect line. The short answer is that the lm() procedure cannot usually do a "perfect" job of separating the predictors and the noise. Consider that the influence of the three independent variables mixed together with the random noise variable would only be "pure" or "perfect" if all four of those variables were entirely uncorrelated with each

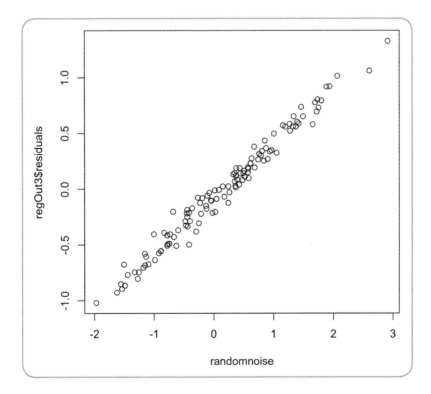

FIGURE 8.6. Residuals from the regression model plotted against the original noise variable.

other. As the following correlation matrix shows, they are all slightly correlated with each other:

```
cor(educdata)
                hardwork     basicsmarts      curiosity    randomnoise
hardwork       1.0000000      0.21591936   -0.135652810    0.139166419
basicsmarts    0.2159194      1.00000000   -0.173518298    0.077928061
curiosity     -0.1356528     -0.17351830    1.000000000   -0.003057987
randomnoise    0.1391664      0.07792806   -0.003057987    1.000000000
```

In fact, just by chance we have one correlation of nearly $r = .22$ and another of $r = -.17$, simply as a result of sampling error. See if you can spot them and name the two variables involved in each correlation. Whenever two independent variables are correlated with each other, the slope, or "B-weight," that is calculated for each one only reflects its *independent* influence on the dependent variable. The other part of the influence on the dependent variable is not "lost," per se, but it is also not visible in the coefficients. (In this example, we have the added complication that the random noise component we added to gpa is actually slightly correlated with two of the three independent variables.) The lm() method we used to create this model is quite capable of taking this lack of independence into account, but in extreme cases this can lead to a model with a significant multiple R-squared, yet where all of the B-weights are nonsignificant.

The "car" (companion to applied regression) package provides a range of useful diagnostics for multiple regression models, including a test for **multicollinearity**—situations where the independent variables are so highly correlated with one another that the model results are potentially inaccurate. If you spend some time exploring the car package, try checking out the vif() function as a diagnostic for multicollinearity.

THE BAYESIAN APPROACH TO LINEAR REGRESSION

As before, we can use a Bayesian approach to inference to learn more about the likelihoods surrounding each parameter of our model. In effect, the Bayesian approach gives us direct and detailed probability information about each "alternative hypothesis" instead of just the "null hypothesis." For each Bayesian analysis, then, we start with assumptions about priors, modifying our prior beliefs with information from our data set, and ending up with posterior probability distributions for each coefficient. Rouder and Morey (2012) documented their reasoning about priors for both univariate and multiple regression. In both cases, the Cauchy distribution (which looks like a normal distribution with heavier tails) simplifies the computation of the Bayes factor and provides satisfactory results under a wide range of conditions. The Cauchy distribution is used to model the standardized version of the B-weight, known as "beta." Rouder and

Morey (2012) suggest setting the standard deviation of the Cauchy priors to 1, which translates into a prior belief that beta will generally fall in the range of −1 to +1 with values in the central region close to zero being more likely (think of a normal distribution centered on 0). They characterize this use of the Cauchy prior with a standard deviation of 1 as "weakly informative," because it represents diffuse knowledge about the likelihood of various beta values.

As with the ANOVA analysis, we will use the BayesFactor package, which offers a specialized function, lmBF(), to conduct the linear multiple regression analysis. Recall that the basic syntax for specifying a model is the same as for the regular lm() function:

```
regOutMCMC <- lmBF(gpa ~ hardwork + basicsmarts + curiosity,
    data=educdata, posterior=TRUE, iterations=10000)
summary(regOutMCMC)
```

In the commands above, we run lmBF() with posterior=TRUE and iterations=10000 to sample from the posterior distribution using the Markov chain Monte Carlo (MCMC) technique. Looking at the MCMC output first, we see both the means of the respective distributions and the 95% HDIs:

1. Empirical mean and standard deviation for each variable, plus standard error of the mean:

	Mean	SD	Naive SE	Time-series SE
hardwork	0.5629	0.05086	0.0005086	0.0005086
basicsmarts	0.5220	0.05027	0.0005027	0.0004712
curiosity	0.5057	0.04394	0.0004394	0.0004394
sig2	0.2545	0.03391	0.0003391	0.0003715

2. Quantiles for each variable:

	2.5%	25%	50%	75%	97.5%
hardwork	0.4625	0.5295	0.5629	0.5964	0.6633
basicsmarts	0.4238	0.4883	0.5216	0.5556	0.6217
curiosity	0.4192	0.4768	0.5054	0.5355	0.5908
sig2	0.1968	0.2309	0.2516	0.2749	0.3308

In the output displayed above, we have parameter estimates for the B-weights of each of our predictions (the column labeled "Mean"). You will notice that the values are quite similar, but not identical to what was produced by the lm() function. In the second section, we have the 2.5% and 97.5% boundaries of the HDI for each of the B-weights. As with previous considerations of the HDI, these boundaries mark the edges of the central region of the posterior distribution for each B-weight. Notably, the distributions for all three predictors are quite similar, which is to be expected given how we constructed these made up variables. We can take a closer look at one of the HDIs by examining a histogram that displays the distribution of B-weights for hardwork across all 10,000 posterior samples as shown in Figure 8.7:

```
hist(regOutMCMC[,"hardwork"])
abline(v=quantile(regOutMCMC[,"hardwork"],c(0.025)), col="black")
abline(v=quantile(regOutMCMC[,"hardwork"],c(0.975)), col="black")
```

The histogram in Figure 8.7 confirms that the estimates are symmetrically distributed centered on 0.56. We have used quantile() and abline() to add vertical lines to this diagram that graphically show the extent of the 95% HDI. One other aspect of the output to notice is the "sig2" estimates summarized in the lmBF() output. The sig2 abbreviation refers to a "model precision" parameter for each of the 10,000 iterations (sig2 is an abbreviation for sigma-squared). This sig2 parameter summarizes the error in the model: the smaller sig2 is, the better the quality of our prediction. With a little math we can use these sig2 estimates to calculate and display the mean value of R-squared, as well as a distribution of R-squared values. The R-squared for each model in the posterior distribution is equal to 1 minus the value of sig2 divided by the variance of the dependent variable. The following code calculates a list of these R-squared values—one for each of our 10,000 posterior samples—displays the mean value, and shows a histogram in Figure 8.8.

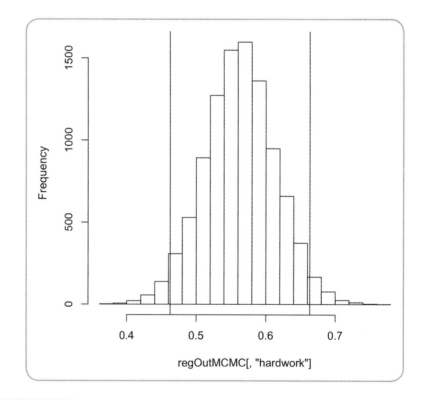

FIGURE 8.7. Distribution of Bayesian estimates of the B-weight on hardwork with vertical lines marking the 95% HDI.

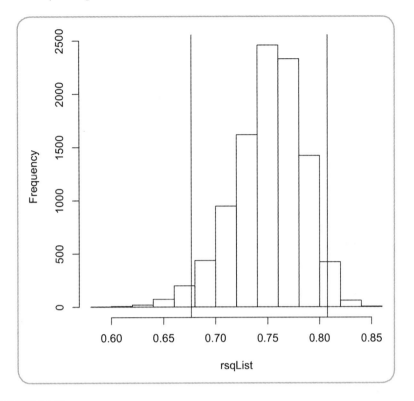

FIGURE 8.8. Distribution of Bayesian estimates of *R*-squared, with vertical lines marking the 95% HDI.

```
rsqList <- 1 - (regOutMCMC[,"sig2"] / var(gpa))
mean(rsqList)            # Overall mean R-squared is 0.75
hist(rsqList)            # Show a histogram
# Lower bound of the 95% HDI
abline(v=quantile(rsqList,c(0.025)), col="black")
# Upper bound of the 95% HDI
abline(v=quantile(rsqList,c(0.975)), col="black")
```

Note that the mean value of this distribution came out to 0.75, which is slightly lower than the *R*-squared that we obtained from the lm() model but almost precisely equal to the adjusted *R*-squared we obtained from that model. While this will not always precisely be the case, the Bayesian model does give us a clear-eyed view of the likely range of possibilities for the predictive strength of our model. In the underlying population that this sample represents, it is credible for us to expect an *R*-squared as low as about 0.67 or as high as about 0.81, with the most likely values of *R*-squared in that central region surrounding 0.75.

By the way, can you reason out why the distribution is asymmetric with a skew to the left? Remember that *R*-squared ranges from 0 to 1. By definition an R-squared of 0 indicates that there is no variance in common between the predictors (as a set) and the dependent variable. (That is also the definition of the null hypothesis.) On the other hand, an R-squared of 1 indicates that the dependent variable is perfectly predicted by the set of predictors (independent variables). You should be able to see that as R-squared gets up close to 1 (as it is here with values as high as 0.85), there is a ceiling effect that squashes the right-hand tail. Of course, the fact that the distribution of R-squared values is asymmetric here makes it extra important that you plot the 95% HDI lines so that you can spot where the most likely values of R-squared lie.

Finally, we can obtain the Bayes factor for our model, although with the weight of all of the evidence we have gathered so far, we would certainly expect a very strong Bayes factor:

```
regOutBF <- lmBF(gpa ~ hardwork + basicsmarts + curiosity, data=educdata)
regOutBF
```

That final command displays the following output on the console:

```
Bayes factor analysis
--------------
[1] hardwork + basicsmarts + curiosity : 7.885849e+32 ±0%
Against denominator:
Intercept only
---
Bayes factor type: BFlinearModel, JZS
```

This shows that the odds are overwhelmingly in favor of the alternative hypothesis, in the sense that a model containing hardwork, basicsmarts, and curiosity as predictors is hugely favored over a model that only contains the Y-intercept (in effect, the Y-intercept-only model means that you are forcing all of the B-weights on the predictors to be 0). Remember that the e+32 is scientific notation: move the decimal point 32 places to the right if you want to see that giant number. Also remember that the Bayes factor is an odds ratio showing the likelihood of the stated alternative hypothesis (in this case a model with nonzero weights for the predictors) divided by the likelihood of the null model (in this case the intercept-only model).

A LINEAR REGRESSION MODEL WITH REAL DATA

Before we close this chapter, let's examine, analyze, and interpret regression results from a real data set. The built-in data sets of R contain a data matrix called state.x77 that has eight different statistics pertaining to the 50 U.S. states. The statistics were obtained from U.S. Census data, mainly in the 1970s. We will use a variable representing life expectancy as our dependent variable. We will try to predict life expectancy using three predictors: the percentage of high school graduates, the per capita income, and the percentage of the population that is illiterate. You should not have to run data(state.x77), because R loads these built-in data by default. The state.x77 data object is actually a matrix, so let's first convert it to a data frame to make it easier to work with:

```
stateData <- data.frame(state.x77)
```

I encourage you to examine the contents of this data frame with com-mands such as str(), dim(), summary(), and cor(). These commands will help to orient you to the eight variables in this data set in order to prepare for the following commands:

```
stateOut <- lm(Life.Exp ~ HS.Grad + Income + Illiteracy,data=stateData)
summary(stateOut)
```

I will not reproduce the complete output of the summary() command above, but just show you the key elements describing the results:

Coefficients:

	Estimate	Std. Error	t value	Pr(>ltl)
(Intercept)	69.0134837	1.7413602	39.632	<2e-16 ***
HS.Grad	0.0621673	0.0285354	2.179	0.0345 *
Income	-0.0001118	0.0003143	-0.356	0.7237
Illiteracy	-0.8038987	0.3298756	-2.437	0.0187 *

Residual standard error: 1.06 on 46 degrees of freedom
Multiple R-squared: 0.4152, Adjusted R-squared: 0.377
F-statistic: 10.89 on 3 and 46 DF, p-value: 1.597e-05

Let's work our way up from the bottom this time: the overall R-squared is 0.4152. The null hypothesis test on this R-squared—which asserts that R-squared is actually 0 in the population—has $F(3,46) = 10.89$, $p < .001$, meaning that we reject the null hypothesis. If the null hypothesis were true in the population, the likelihood of observing a value of F greater than 10.89 is extremely small (specifically $p = 0.00001597$). This value of p is well below all of the conventional alpha thresholds of 0.05, 0.01, and 0.001. The adjusted R-squared of 0.377 shows a bit of shrinkage as a result of using three predictors with a small sample. Nonetheless, a significant R-squared suggests that at least one of our predictors is accounting for some of the variance in the dependent variable, Life.Exp.

The coefficients in the "Estimate" column show the B-weights for each predictor: a positive weight for HS.Grad, a weight near zero for Income, and a negative weight for illiteracy. The t-tests and corresponding p-values in the right-most two columns in each case test the null hypothesis that the coefficient equals 0 in the population. Both HS.Grad and Illiteracy are significant, because they have p-values less than the conventional alpha of $p < .05$. Thus, we can reject the null hypothesis in both of those cases. The p-value for the t-test on Income, 0.7237, is higher than $p < .05$, so in that case we fail to reject the null hypothesis. Don't forget that failing to reject the null does not mean that the null is true. We have not shown statistically that the B-weight on Income is zero. Put this all together and the results suggest that the higher the percentage of high school graduates in a state, the higher the life expectancy and the lower the illiteracy rate, the higher the life expectancy. Both results root for the value of education!

We can conduct a Bayesian analysis of the same model using the following code:

```
stateOutMCMC <- lmBF(Life.Exp ~ HS.Grad + Income + Illiteracy,
     data=stateData, posterior=TRUE, iterations=10000)
summary(stateOutMCMC)
rsqList <- 1 - (stateOutMCMC[,"sig2"] / var(stateData$Life.Exp))
mean(rsqList)            # Overall mean R-squared
quantile(rsqList,c(0.025))
quantile(rsqList,c(0.975))
```

As demonstrated earlier in this chapter, we use the same model specification to call lmBF(), with the addition of the posterior=TRUE and iterations=10000 arguments to obtain the posterior distributions of each parameter. The results provided by the summary() command provide several similarities to the conventional analysis and a couple of striking differences. The mean B-weights for HS.Grad and Income are very similar to what the conventional analysis showed. The 95% HDI for the B-weight of income overlaps with 0, providing evidence that the population value of that B-weight does not credibly differ from 0.

In contrast, the mean of the posterior distribution of R-squared values is 0.30, quite a bit smaller than the adjusted R-squared of 0.377 for the conventional analysis. Note that the 95% HDI for the posterior distribution of R-squared values ranges from 0.01 to 0.57. The reason that the mean R-squared is so much lower in the Bayesian analysis is that the mean B-weight on Illiteracy is also smaller in magnitude for the Bayesian analysis than for the conventional analysis, −0.71 as opposed to −0.80. In other words, the Bayesian analysis finds that Illiteracy is not as strong a predictor as the conventional analysis indicated.

The reason for this difference may be that the Illiteracy variable is highly positively skewed, which I learned by generating a histogram of the Illiteracy variable with the command hist(stateData$Illiteracy). In the vast majority of states, illiteracy rates were very low when these data were collected, but for a small number of states illiteracy rates were quite a bit higher. This creates a strong right tail on the distribution of the Illiteracy variable. The conventional multiple regression, using the least-squares fitting method, expects each predictor and the dependent variable to be normally distributed, whereas Illiteracy is quite nonnormal because of that positive skew. The violation of the normality assumption may be inflating the B-weight for Illiteracy in the conventional analysis.

As a result of that skewness, the Bayesian analysis is finding a few posterior parameter configurations where the B-weight on Illiteracy is notably distant from the mean of -0.71. You can verify this for yourself by examining a histogram of the posterior distribution of B-weights with hist(stateOutMCMC[,"Illiteracy"]) or a boxplot of the posterior distribution using boxplot(as.numeric (stateOutMCMC[,"Illiteracy"])). As an experiment, you may wish to increase the number of posterior samples beyond 10,000 to see whether and how the results may differ with a more comprehensive posterior distribution. The discrepancy between the coefficients for the conventional analysis and for the Bayesian analysis clearly demonstrates how important it is to look at your data from multiple angles and to avoid accepting the authority of any one method of statistical inference.

I also ran the following code to obtain a Bayes factor for the overall model:

```
stateOutBF <- lmBF(Life.Exp ~ HS.Grad + Income + Illiteracy,data=stateData)
stateOutBF
```

The output from that second command showed a Bayes factor of 1467.725, very strong positive evidence in favor of the alternative hypothesis. In this case the alternative hypothesis is that the B-weights on HS.Grad, Income, and Illiteracy are nonzero. Taken together, the results of the conventional analysis and the Bayesian analysis indicate that we do have a solid predictive model, where the rate of high school graduation and the illiteracy rate each independently predict life expectancy across the 50 states. The results of the Bayesian analysis also suggested that the B-weight on income was not credibly different from 0, so it does not appear to predict life expectancy independently of the other predictors. Finally, although we have some conflicting evidence on the exact magnitude of R-squared and the B-weight on Illiteracy, the 95% HDIs provide meaningful evidence of a credible difference from 0 for both of these coefficients.

CONCLUSION

In this chapter, we examined linear regression, also called multiple regression, in situations where there is more than one predictor. Multiple regression is an example, one might say one of the best examples, of the general linear method (GLM). With GLM we usually have one outcome variable (sometimes referred to as a dependent variable), which is modeled as a function of a set of predictors (often called independent variables). At the beginning of this chapter we considered some metric predictors that were generated randomly from the standard normal distribution. Multiple regression can also work with predictors that are categorical. In fact, if you are a really curious person, you could repeat the analyses that we did in the ANOVA chapter using lm() and lmBF() in order to compare your results from the ANOVA analysis to what we have accomplished in this chapter using multiple regression. Just change the call to aov() into a call to lm().

As with the previous chapters, we examined both the traditional way of conducting multiple regression analysis as well as an approach using Bayesian methods and reasoning. With the traditional analysis, we demonstrated the use of the least-squares criterion to minimize the sum-of-the-squared errors of prediction. Based on the standard frequentist reasoning, this method allows us to test the null hypothesis that each regression coefficient is 0 in the population as well as the null hypothesis that R-squared (the overall goodness of the model) is also equal to 0 in the population. When predictors correlate strongly enough with the outcome variable, we can often reject these null hypotheses to imply support for the alternative hypotheses.

Using Bayesian methods and reasoning, we can zoom in more closely on the alternative hypotheses by examining the characteristics of the posterior distribution. The information we have about our outcome and our predictors allows us to generate a distribution for each regression coefficient (B-weight) as well as for the value of R-squared. We can look to the center of these distributions to get a sense of where each coefficient value lies and we can also construct a highest density interval (HDI) to get a sense of the range of values that are possible given the vagaries of sampling.

Both the traditional and Bayesian methods give us a couple of different ways of using regression results. Most researchers are concerned with finding out whether a particular independent variable works in providing a meaningful connection to a dependent variable. Researchers sometimes reason about the relative strength of different predictors or they compare models with different predictors. For practical applications of regression results, the coefficients are sometimes used to create a forecasting model where new data about predictors can be used to predict a hypothetical outcome. If we were designing a new model of automobile and we wanted an educated guess as to the fuel efficiency of the new model, we might forecast the mpg based on the proposed weight of the new vehicle and its engine size. Either way, multiple regression is one of the workhorse techniques used most frequently in applied statistics.

There's much more to know about multiple regression that we did not cover in this chapter, especially around diagnostics and proper interpretation of models, but there is one final, important concept that you should know before you turn to the exercises. In the made-up educational data example that we used early in the chapter, we constructed and used some standard normal variables. This made it easy to see, for example, that each of our three predictors was contributing about the same amount of explanatory power to our prediction model. When we started to work with real data, you may have noticed that the B-weight on each predictor reflected not just the strength of prediction but also the scale on which the different variables were measured. After all, if you are predicting the weights of some objects in kilograms, the coefficients in the prediction equations are going to be very different than if you are predicting weights in ounces. Fortunately, there are useful ways of standardizing B-weights and also of comparing these standardized weights to help with an understanding of which predictors are important and which predictors less so.

EXERCISES

1. The data sets package in R contains a small data set called mtcars that contains $n = 32$ observations of the characteristics of different automobiles. Create a new data frame from part of this data set using this command: myCars <- data.frame(mtcars[,1:6]).

2. Create and interpret a bivariate correlation matrix using cor(myCars) keeping in mind the idea that you will be trying to predict the mpg variable. Which other variable might be the single best predictor of mpg?

3. Run a multiple regression analysis on the myCars data with lm(), using mpg as the dependent variable and wt (weight) and hp (horsepower) as the predictors. Make sure to say whether or not the overall R-squared was significant. If it was significant, report the value and say in your own words whether it seems like a strong result or not. Review the significance tests on the coefficients (B-weights). For each one that was significant, report its value and say in your own words whether it seems like a strong result or not.

4. Using the results of the analysis from Exercise 2, construct a prediction equation for mpg using all three of the coefficients from the analysis (the intercept along with the two B-weights). Pretend that an automobile designer has asked you to predict the mpg for a car with 110 horsepower and a weight of 3 tons. Show your calculation and the resulting value of mpg.

5. Run a multiple regression analysis on the myCars data with lmBF(), using mpg as the dependent variable and wt (weight) and hp (horsepower) as the predictors. Interpret the resulting Bayes factor in terms of the odds in favor of the alternative hypothesis. If you did Exercise 2, do these results strengthen or weaken your conclusions?

6. Run lmBF() with the same model as for Exercise 4, but with the options posterior=TRUE and iterations=10000. Interpret the resulting information about the coefficients.

7. Run install.packages() and library() for the "car" package. The car package is "companion to applied regression" rather than more data about automobiles. Read the help file for the vif() procedure and then look up more information online about how to interpret the results. Then write down in your own words a "rule of thumb" for interpreting vif.

8. Run vif() on the results of the model from Exercise 2. Interpret the results. Then run a model that predicts mpg from all five of the predictors in myCars. Run vif() on those results and interpret what you find.

9. The car package that you loaded for Exercise 6 contains several additional data sets that are suitable for multiple regression analysis. Run data(Duncan) to load the Duncan data on the prestige of occupations. Analyze the hypothesis that the "income" variable and the "education" variable predict the "prestige" variable. Make sure to use both conventional and Bayesian analysis techniques and to write an interpretation of your results that integrates all of the evidence you have gathered.

10. Bonus Statistical Thinking Exercise: Here is a block of code that does something similar to what we did at the beginning of the chapter with the made-up GPA data set:

```
set.seed(1)
betaVar <- scale(rbeta(50,shape1=1,shape2=10))
normVar <- rnorm(50)
poisVar <- scale(rpois(50,lambda=10))
noiseVar <- scale(runi(50))
depVar <- betaVar/2 + normVar/2 + poisVar/2 + noiseVar/2
oddData <- data.frame(depVar,betaVar,normVar,poisVar)
```

You can see by that final command that we are creating another made-up data set that is appropriately named "oddData," as it consists of variables from several different distributions. Explain what each line of code does. Explain why this data set causes some difficulties with conventional multiple regression analysis. Cut/paste or type in the code to create the data set, then run lm() on the data, using depVar as the dependent variable and betaVar, normVar, and poisVar as the predictors. Interpret the results. For a super bonus, run lmBF() on the data and also interpret those results.

CHAPTER 9

Interactions in ANOVA and Regression

In earlier chapters we have examined some of the most common configurations of the general linear model, namely, the analysis of variance and linear multiple regression. Across the board, we examined simple, direct effects of the independent variables of interest on the dependent variable. Statisticians often refer to these as *main effects*. So when we say that hardwork predicts gpa or that the type of transmission in an automobile affects its fuel economy, we are describing main effects.

When there are two or more main effects at work in a model, however, there is an additional possibility called an **interaction.** The basic idea of an interaction is that the dependent variable is affected differently by one independent variable depending upon the status of another independent variable. The example we will use to illustrate this first comes from an R package called HSAUR. The HSAUR package contains a data set called "weightgain" which contains two predictor variables and an outcome variable. The study that originated these data tested the amount of weight gain by laboratory rats fed two different diets, one of meat (beef) or one of cereal. Feeding beef to rats: Weird! In addition, rats in both conditions either received a high-protein diet or a low-protein diet. Thus, overall, there were four conditions determined by these two independent variables:

- Beef/high protein
- Beef/low protein
- Cereal/high protein
- Cereal/low protein

The research question thus has three parts:

1. Did beef versus cereal make a difference to weight gain?
2. Did low versus high protein make a difference to weight gain?
3. Was the difference between low and high protein greater for the beef diet than for the cereal diet?

The first two questions are questions about main effects and the third question is about an interaction. It is important that you grasp the difference in the ways that the first two questions were posed in contrast to the third question. To answer the third question, one needs to simultaneously consider the status of the beef/cereal factor as well as the high/low protein factor. By the way, it would be fine to phrase the interaction research question the other way: Was the difference between beef and cereal greater for the high-protein diet than for the low-protein diet? Although the questions seem different, the same statistical test sheds light on either one.

Interactions are often easier to understand from a diagram than from a verbal description. To get an initial view of the situation we can run a useful function called interaction.plot() to visualize the data in our weight-gain data set as shown in Figure 9.1.

```
install.packages("HSAUR")
library("HSAUR")
data("weightgain", package = "HSAUR")


# Create and display an interaction plot
wg <- weightgain   # Copy the dataset into a new object
interaction.plot(x.factor=wg$source,trace.factor=wg$type,
    response=wg$weightgain)
```

In the diagram in Figure 9.1, the Y-axis represents the amount of weight gain in grams (remember, these are rats). The X-axis notes two different categories of food source: on the left we have beef and on the right we have cereal. Finally, we have two different lines, one representing the type of diet: high protein or low protein. The fact that there are lines in this diagram is a bit weird: The endpoints of the lines represent the means for the four different groups. None of the intermediate points implied by the lines really exist—for example, there is nothing "in between" beef and cereal. So the lines are simply there to help you visually connect the two means that belong to the same condition of the second factor (type: high vs. low).

Look closely at the two lines. Your primary clue that there may be an interaction present is that the two lines are not parallel. That suggests the possibility that the effect of food source on weight gain may be dependent upon (i.e., it interacts with) whether the diet type is high protein or low protein. In fact, if you look at the specific position of the four endpoints of the lines you should be able to conclude as follows: Providing a high-protein beef diet creates a substantial gain over a low-protein beef diet, whereas providing a high-protein cereal diet only provides a small gain over a low-protein cereal diet. That previous sentence represents the interaction, if one is present. Make a habit of inspecting an interaction diagram before drawing any conclusions about the data: when the lines are not parallel, that is our hint that an interaction effect may exist. The phrase "may exist" is very important: we need inferential reasoning to satisfy ourselves that the differences we are seeing are not simply due to sampling error.

This is where the ANOVA test comes back into play. Earlier in the book we tested what statisticians call **oneway ANOVA**. That means that there was just a single factor with two or more levels. Oneway ANOVA is particularly well suited to testing a single factor that has more than two levels. For example, one might compare sales of books in three different formats, hard cover, paperback, and eBook. Or one might compare the level of philosophy knowledge among students from four different majors: English, Sociology, Mathematics, and Physics. Here in this chapter we are now making a transition to **factorial ANOVA**, where we have at least two factors (i.e., at least two independent variables) and each factor has two or more levels.

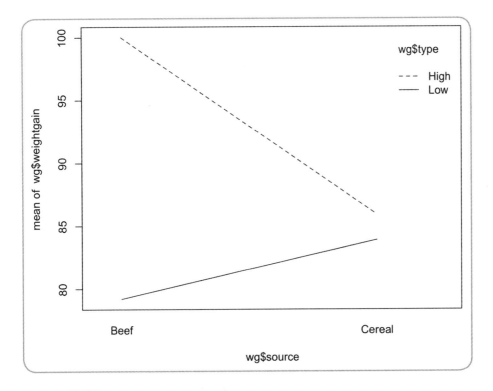

FIGURE 9.1. Interaction plot of weightgain (source versus type).

INTERACTIONS IN ANOVA

Looking back to Figure 9.1, you can see that this is the basic arrangement that we have in our weight-gain data set: a dietary source factor with two levels (beef, cereal) fully crossed with a type factor that also has two levels (low protein and high protein). The phrase "fully crossed" means that for each level of the first factor we have all levels of the second factor (and vice versa). Either of the following two aov() calls will test the main effects of these two factors as well as the interaction between them. Both calls run the identical analysis and provide the same output:

```
aovOut = aov(weightgain ~ source + type + source:type, data=weightgain)
aovOut2 = aov(weightgain ~ source * type, data=weightgain)
```

The model specification in the first version of the command explicitly lists all of the effects we want to test: the main effect of source, the main effect of type, and the interaction between source and type ("source:type"). The model specification in the second command contains a shortcut in the formula that accomplishes the same thing with less verbiage. Let's take a look at the output:

```
summary(aovOut2)
             Df  Sum Sq  Mean Sq  F value   Pr(>F)
source        1     221    220.9    0.988   0.3269
type          1    1300   1299.6    5.812   0.0211 *
source:type   1     884    883.6    3.952   0.0545 .
Residuals    36    8049    223.6
---
Signif. codes: 0 '***' 0.001 '**' 0.01 '*' 0.05 '.' 0.1 ' ' 1
```

Most statisticians recommend examining and interpreting the interaction before the main effects, because the interpretation of the interaction may supersede the interpretation of the main effects. In this case, you can see that the $F(1,36) = 3.95$ is *not* statistically significant at the conventional alpha threshold of $p < .05$. Thus, despite the fact that the interaction plot in Figure 9.1 seemed promising, *there is no statistically significant interaction*. Now is an important time to reinforce our strategy for the interpretation of p-values: The fact that this observed p-value, 0.0545, only exceeds the $p < .05$ threshold by a tiny amount is of no consequence. Any p-value greater than or equal to 0.05 is not statistically significant, end of story. Reread the end of Chapter 5 if you need a reminder of why we should be sticklers about this point.

Now that we have interpreted the significance of the interaction, and found it not significant, we can move along to interpreting the main effects. No significant effect appeared for the main effect of source. Some researchers would report this as $F(1,26) = 0.988$, N.S., where the notation "N.S." means

Degrees of Freedom for Interactions

The ANOVA output for our test of rat weight gain shows $F(1,36) = 3.952$, $p = .0545$, for the null hypothesis test of the interaction term. But why does the interaction have 1 degree of freedom between groups and 36 degrees of freedom within groups?

Remember that the F-test references a family of distributions that varies somewhat in shape based on two parameters: between-groups degrees of freedom and within-groups degrees of freedom. So for every F-test we conduct—and there are three of them in this ANOVA analysis—we need to know the between-groups degrees of freedom and the within-groups degrees of freedom.

This data set has 40 observations in it, which is the starting point for our calculation of degrees of freedom. Once the grand mean is calculated—it is the basis of all of our later variance calculations for the ANOVA—there are only 39 degrees of freedom left. Each of our main effects has two levels, and the degrees of freedom for a given factor is always 1 less than the number of levels in that factor $(k - 1)$, so 1 degree of freedom is lost for dietary source (beef vs. cereal) and another is lost for type (high vs. low protein), leaving 37.

The degrees of freedom taken by an interaction term is the *product of the degrees of freedom for the main effects that go into it*, in this case $1 * 1 = 1$. In our rat weight gain study, that leaves exactly 36 degrees of freedom: this is the number of residuals that are free to vary after everything else has been calculated. So $df = 36$ is the within-groups degrees of freedom.

Let's practice with one additional example of a study with a different setup: 60 participants (half men and half women) were randomly assigned to one of three gaming conditions: puzzle, adventure, and strategy. The gender factor has $k = 2$ levels: male and female. The game type factor has $k = 3$ levels: puzzle, adventure, and strategy. Start with 60 degrees of freedom and subtract 1 for the grand mean, leaving 59. The between-groups degrees of freedom for gender is $2 - 1 = 1$. That leaves 58 degrees of freedom. The between-groups degrees of freedom for game type is $3 - 1 = 2$. That leaves 56 degrees of freedom. The degrees of freedom for the interaction term is $1 * 2 = 2$, leaving 54 degrees of freedom within groups. So the F-test for the main effect of gender would be on $F(1,54)$ degrees of freedom, whereas the F-test for game type would be $F(2,54)$. The F-test for the interaction would also be on $F(2,54)$ degrees of freedom.

Don't forget that you can explore the shapes of the family of F-distributions yourself with the rf(n, df1, df2) function. For example, try these commands to empirically simulate the F-distribution for 1 and 36 degrees of freedom:

```
testF <- rf(n=10000, df1=1, df2=36)     # Generate random Fs for F(1,36)
hist(testF)                             # Display a histogram
```

(continued)

```
# Show threshold of significance, p <.05
abline(v=quantile(testF,c(0.95)),col="black")
quantile(testF,c(0.95))                          # Report the threshold to the console
```

Remember that the randomly generated values of *F* represent the distribution of outcomes under the assumption of the null hypothesis for the particular combination of *df* values you are exploring. Thus, as is the case with other test statistics, we are looking for the presence of extreme values of the test statistic as a way of rejecting the null hypothesis. The code above places a vertical line at the point that divides the *F* distribution into a central region on the left side containing 95% of the points and a tail on the right side containing 5% of the points. That vertical line represents the threshold of statistical significance (assuming that you choose a conventional alpha of $p < .05$).

not significant. A significant main effect appeared for the main effect of type, with $F(1,36) = 5.81$, $p < .05$.

Let's now rerun the analysis using the BayesFactor package to see if the Bayesian evidence fits with the evidence from the null hypothesis tests:

```
aovOut3 = anovaBF(weightgain ~ source*type, data=weightgain)
aovOut3
```

That code produces the following output:

```
Bayes factor analysis
--------------
[1] source                      : 0.4275483 ±0%
[2] type                        : 2.442128 ±0%
[3] source + type               : 1.037247 ±1.1%
[4] source + type + source:type : 1.724994 ±0.88%
Against denominator: Intercept only
```

These results confirm some of our earlier findings. First, source has a Bayes factor less than 1, so that strongly supports the absence of an effect, also known as the null hypothesis. To get a sense of support for the null hypothesis, you can invert the fractional odds ratio: 1/0.428 = 2.34. This provides only very weak evidence in favor of the null hypothesis of no effect for source. Next, the type factor has an odds ratio of 2.44:1 in favor of an effect for type. Remember that according to the rule of thumb provided by Kass and Raftery (1995; see "Interpreting Bayes Factors" in Chapter 6), any odds ratio between 1:1 and 3:1 is "barely worth mentioning." In other words, the odds in favor of the alternative hypothesis for type, while better than even, are quite weak. The next line of output contains a model that aov() did not consider, that is, a main effects-only model. This has an odds ratio of about 1:1, so it too is unworthy of further consideration. Finally, the full model containing the interaction term has an odds ratio of less than 2:1; this is extremely weak evidence in favor of the interaction effect.

We should take care to compare the interaction model versus the main effects-only model. Even though we know in this case that both models have very weak support, in the more general situation we want to examine whether the model that includes the interaction is noticeably better than the main effects-only model. The BayesFactor package makes it easy for us to do this. Look in the output just above and you will see that each of the four models has an index next to it: [1], [2], [3], and [4]. We can use these indices to compare nested Bayesian models, simply by creating a fraction or ratio of the two models, like this:

```
aovOut3[4]/aovOut3[3] # What's the odds ratio of model 4 vs. model 3?
```

Typing that expression at the command line produces the following output:

```
Bayes factor analysis
--------------
[1] source + type + source:type : 1.576853 ±1.15%
Against denominator: weightgain ~ source + type
```

This result shows odds of 1.6 to 1 in favor of the model that includes the interaction term. Again following the rules of thumb we examined in "Interpreting Bayes Factors" in Chapter 6, this would also be considered a very weak result, barely worth mentioning. Please note that because the BayesFactor package is using MCMC to explore the posterior distributions, your results from the anovaBF() test may not precisely match what I have displayed above. The sample size is small enough in this data set that the random numbers used to control the posterior sampling will have some effect on the results. If you use set.seed() prior to the anovaBF() call, you can stabilize the output.

We can examine the posterior distribution of key parameters in a similar fashion to what we accomplished for oneway ANOVA. Remember that anovaBF() models each parameter as a deviation from the population grand mean, labeled as "mu" in the detailed output. Because the detailed output is quite lengthy, I summarize the results of the posterior sampling with box plots in Figures 9.2 and 9.3:

```
mcmcOut <- posterior(aovOut3[4],iterations=10000)   # Run mcmc iterations
boxplot(as.matrix(mcmcOut[,2:5]))                    # Figure 9.2
boxplot(as.matrix(mcmcOut[,6:7]))                    # Figure 9.3
```

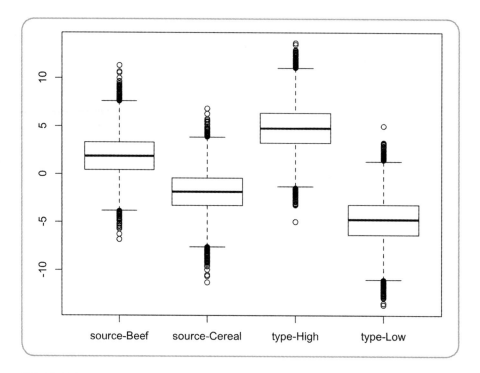

FIGURE 9.2. Box plot of posterior distributions for deviations from mu (the grand mean) for the source main effect and the type main effect.

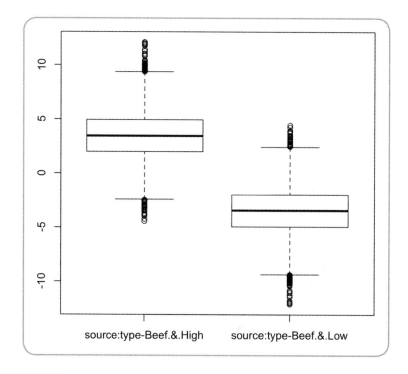

FIGURE 9.3. Box plot of posterior distributions for deviations from mu (the grand mean) for one of the interaction contrasts. Beef&High versus Beef&Low.

The first box plot, which displays the main effects, appears in Figure 9.2. When examining Figure 9.2, keep in mind that the leftmost two box plots, representing the main effect of beef versus cereal, ignores or "collapses across" the high- versus low-protein condition. Likewise the rightmost two box plots, comparing high versus low protein, ignores or "collapses across" the beef versus cereal condition. So in examining main effects we should examine the leftmost box plots as a pair and the rightmost box plots as a pair. The whiskers on each box cover the full range of the distribution except for a few outliers, and as such can be considered roughly equivalent to the 95% HDI. The leftmost pair of boxes overlap substantially, confirming our previous understanding that the source factor (beef vs. cereal) has no credible difference between the two levels. Looking at the rightmost pair of boxes/whiskers, we can also see that the low-protein and high-protein box plots also overlap to some degree. Putting this evidence in context, although the significance test showed that the main effect of type (high vs. low protein) was statistically significant, the lack of complete separation of the box plots confirms what the weak Bayes factor (2.44) showed: the main effect of type is so weak as to be barely worth mentioning.

One of the interaction contrasts, showing "Beef&High" versus "Beef&Low," appears in Figure 9.3. You can interpret Figure 9.3 in a similar way to Figure 9.2: The whiskers roughly represent the edges of the 95% HDI, and for a credible difference we would expect minimal or preferably no overlap between the bottom whisker of Beef&High and the top whisker of Beef&Low. Note that Figure 9.3 only represents one of the two contrasts we would have to examine to definitively interpret a credible interaction. The research question—Was the difference between low and high protein greater for the beef diet than for the cereal diet?—suggests that we would compare Figure 9.3 to the contrast of Cereal&High versus Cereal&Low. Given that the significance test and the Bayes factor both undermine support for the interaction, we need not conduct that comparison in this case.

Given the apparent strength of the interaction that we thought we saw in Figure 9.1, you may be wondering why this data set was such a wash. Let's briefly return to the raw data and reexamine the mean plots with an interesting new diagnostic. The gplots library provides the plotmeans() function, which we can use to make a means plot that shows confidence intervals:

```
install.packages("gplots")
library("gplots")
plotmeans(weightgain ~ interaction(source,type,sep =" "), data = weightgain,
    connect = list(c(1,2),c(3,4)))
```

In Figure 9.2, each mean has a "t"-shaped upper and lower marker, shown in grey, representing plus or minus two standard errors from the mean. As a rule of thumb, plus or minus two standard errors is the same thing as the confidence interval around each mean. Just using your eyeballs you can see the extensive amount of overlap between almost every pair of means. You may want to compare this view of the raw data with Figure 9.1: the two plots show the same information about group means. In both plots the fact that the lines are not parallel suggests the possibility of an interaction. Yet by adding the information about confidence intervals, Figure 9.4 shows more clearly that the pattern of differences in the groups is not large enough to overcome the uncertainty due to sampling error. If you were reporting the results of this analysis, Figure 9.4 would be worthwhile to show alongside the Bayesian and conventional statistical tests.

To conclude our analysis of these hungry rats, we failed to find a significant interaction effect or a significant effect for the source factor. The statistically significant effect for type (using the conventional ANOVA test) was shown by the Bayes analysis to have only the weakest level of support—"barely worth mentioning." Regardless of the fact that the p-value associated with the F-test on the main effect of type was 0.02114 (and therefore significant), we know that the Bayes factor of 2.44:1 means that the odds are barely in favor of the "alternative hypothesis" of an effect for type.

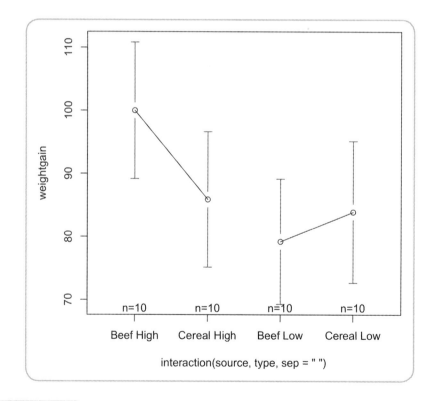

FIGURE 9.4. Means plot of weight-gain data with confidence intervals indicated.

A Word about Standard Error

You have heard and used the term "standard deviation" a lot in this book. Another term, standard error, comes up from time to time, but until now I have not taken the time to formally introduce it. The interesting, and slightly confusing, thing about the standard error is that it is a kind of standard deviation. To really understand the distinction, you must cast your mind back to Chapter 3, where we repeatedly sampled from a population in order to create a sampling distribution of means. Let's briefly think that process through again.

If we had a very large random normal distribution, by definition the mean of that population would be (near) 0 and the standard deviation would be (near) 1. We could repeatedly draw samples of 100 from that population and keep a list of them. You know that this distribution of sample means would show some variability, but by exactly how much? As it turns out, statisticians have used their mathematical techniques to show that the variability of a sampling distribution of means could be calculated as follows:

$$\sigma_{\bar{x}} = \frac{\sigma}{\sqrt{n}}$$

This equation states that the standard error of the mean (shown as the Greek letter sigma, with a subscript of x-bar) is equal to a fraction, with sigma (the population standard deviation) in the numerator and the square root of the sample size in the denominator. So for a population with a standard deviation of 1, where we repeatedly draw samples of size $n = 100$, the standard error would be 0.1. Thus, the standard error is directly related to the population standard deviation. The formula shows quite clearly how the variability of a sampling distribution would squeeze smaller and smaller as you raise the sample size used to create the sampling distribution.

Relatedly, you may have wondered about my comment that "plus or minus two standard errors is about the same as the confidence interval." Remember that when we create a sampling distribution, as long as we draw enough samples we will tend to get a normal distribution of means in that distribution. If you divide the standard normal curve into a central region and two tails, with 95% in the central region and 2.5% in each of the tails, it turns out that the vertical dividing lines go in at −1.959964 and +1.959964. Rounding off, 1.96 is generally close enough to 2.0 for informal discussions of results. Keeping in mind that the standard error represents the variability in our sampling distribution, then plus or minus two standard errors covers the whole central 95% region of the sampling distribution. Because confidence intervals are built around sample means that, in the long run, are normally distributed and centered on the population mean, we can be assured that, in the long run, 95% of confidence intervals will contain the population value.

(continued)

So what about this rule of thumb I suggested, about the overlap between the confidence intervals of two means? If you have means for two different samples of data, and their confidence intervals do not overlap at all, it is quite likely that a statistical test comparing them will be significant. Similarly, the greater the distance of separation between the two confidence intervals, the stronger the Bayes factor will tend to be. Unfortunately the rule is not perfect: if two confidence intervals overlap to some small degree, it is still possible that the difference in means will register as significant (although the Bayes factor will probably be weak).

We can take our thinking about the type factor one step further by calculating the partial eta-squared effect size value for its main effect. Go back to the ANOVA table a few pages ago to find that type has a sum of squares of 1300. Divide this by the total sum of squares $1300/(221 + 1300 + 884 + 8049) = 0.12$. Remember that the interpretation of eta-squared is the proportion of variance in the dependent variable accounted for by the independent variable(s). Here, the type of feed (high protein vs. low protein) accounts for 12% of the variability in weight gain. Is this bad or good? By now, you should recognize that whether it is bad or good depends upon the details of the situation. If the high-protein diet is considerably more expensive, and the weight gain is inconsequential, then even though the main effect of type was statistically significant the practical implication might be that the high-protein diet is not worth it. The weakness of the Bayesian evidence for a main effect of type would support this idea.

By the way, in some research reports, you might occasionally find authors who state that a p-value such as $p = 0.0545$ is "approaching significance." This idea is one that has really bothered the critics of the significance test (e.g., Aguinis et al., 2010; Armstrong & Henson, 2004; Daniel, 1998). The NHST was devised as a simple go/no-go decision: an effect is either significant or it is not. When you discuss results of statistical analyses in writing or verbally, make sure you avoid the slippery slope of saying that something "approaches" significance or is "highly" significant. Also make sure to report the effect sizes from your analyses and confidence intervals when you can get them.

One last thing: Aguinis (1995) and others have commented on the difficulty of having enough **statistical power** to detect interaction effects. Statistical power is a concept rooted in the frequentist thinking of the null hypothesis test and it refers to the likelihood of finding a statistically significant effect if one is present. One of the primary failings of many research articles that try to detect interaction effects is that the study design does not include enough observations, that is, the sample size is too small. If your research focuses on testing interactions, you would be well served by doing a statistical power analysis before collecting your data so that you know the minimum amount of data you will need in order to have a good opportunity to detect an interaction effect. Cohen's (1992) article "A Power Primer" provides a concise, easy-to-understand article for planning the power needs of your study.

INTERACTIONS IN MULTIPLE REGRESSION

We know that the general linear model is the basis of both ANOVA and regression, and we have seen hints from the output of aov() and lm() that suggest the underlying unity of results. Interactions in multiple regression can seem a bit different than interactions in ANOVA because we focus on the slopes of different regression lines rather than a pattern of mean differences, but under the hood there is an essential similarity to interactions in both ANOVA and multiple regression. Let's begin again by taking a look at some diagrams.

For this section of the chapter, I will use one of R's built-in data sets from the "lattice" package, known as "environmental." The data set contains 111 observations across four variables. The research question I am posing has to do with the amount of ozone measured in New York City versus the levels of solar radiation. At ground level, ozone is a pollutant: solar radiation causes ozone to be formed from the by-products of internal combustion engines. I should be able to predict ozone levels in part based on radiation levels, because on sunnier days more ozone should be created from the ionizing radiation of the sun. Generally speaking, greater ozone should be associated with more solar radiation. I also predict, however, that this effect will be much less pronounced when there is a strong wind, as the wind will tend to disperse the ozone. This latter prediction is an interaction prediction, because it predicts a different slope between radiation and ozone depending upon the level of an additional variable, in this case, wind speed.

Let's create a scatterplot of radiation and ozone and then impose two different regression lines on it. I can use lm() and abline() to create the regression lines, but I need to divide up the data set into a high-wind portion and a low-wind portion to accomplish this. The following chunk of code will do the trick:

```
# Show regression lines on a scatterplot of radiation vs. ozone
install.packages("lattice")
library(lattice)
data(environmental)
plot(environmental$radiation,environmental$ozone)


# This grey line shows what happens when we only consider
# the data with wind speeds above the median wind speed
hiWind <- subset(environmental, wind > median(environmental$wind))
hiLmOut <- lm(ozone ~ radiation,data=hiWind)
abline(hiLmOut,col="grey")
# This dotted black line shows what happens when we only consider
# the data with wind speeds at or below the median wind speed
loWind <- subset(environmental, wind <= median(environmental$wind))
loLmOut <- lm(ozone ~ radiation,data=loWind)
abline(loLmOut,col="black",lty=3)
```

It is a nice feature of abline() that it will look inside the regression output object created by lm() to find the necessary data (slope and intercept) to plot the regression line. Notice that I specified a simple regression model where radiation predicts ozone. If you try to create a more complex regression model, abline() will warn you that it can only do the line plot based on the first two coefficients (the *Y*-intercept coefficient and the first slope). If you want the slope and intercept to reflect the influence of multiple predictors, make sure you list as the first predictor following the ~ character in the model specification the independent variable that you want on the *X*-axis of your scatterplot. The resulting graph appears in Figure 9.5.

Note that the slope of the dotted black line (low winds) is steeper than the slope of the grey line (high winds). This suggests the possibility that my interaction hypothesis is correct. When the wind is blowing hard, the differences between cloudier days and sunnier days are not so strongly reflected in differences in ozone. When the wind is weak or absent, more radiation associates more strongly with more ozone.

This is a great moment to reconsider the sticky topic of causality. These data—the environmental data set built into the lattice package of R—were collected as part of an observational study. There was no experimental manipulation—we simply have a number of variables that were collected at roughly the same time on a variety of different days. As a result, all of the analyses we are undertaking here are correlational. When using correlational data we must

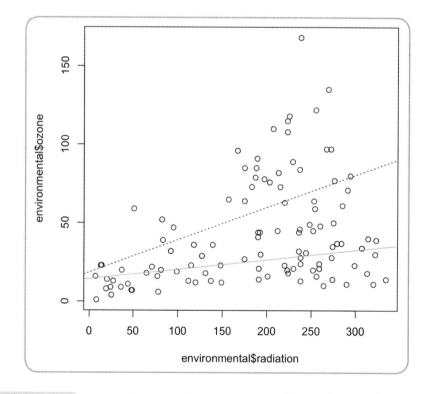

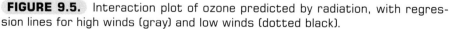

FIGURE 9.5. Interaction plot of ozone predicted by radiation, with regression lines for high winds (gray) and low winds (dotted black).

take care not to make strong statements about the causal relationships among variables, unless we have strong theory to help guide us.

In this particular case, atmospheric scientists can probably make some strong, theory-specific statements about the connections between solar radiation and ground-level ozone. Yet, if we did not have that theory to guide us, there might be plausible alternatives where causality was reversed or a causal connection occurred through an unknown third variable. For instance, on sunny days, the connections between ozone, pollution, and wind may have something to do with how many people go to the beach instead of going to work. We cannot really know for sure because this study did not include any experimental manipulation. If I was an atmospheric scientist, however, and I had good theoretical models of how these variables are connected, I might be able to make stronger statements about the causal connections among them.

With all that in mind, let's now run some inferential tests to see whether there is statistical support for the interaction that the diagram in Figure 9.3 is suggesting. We can use this code:

```
lmOut1 <- lm(ozone ~ radiation * wind, data=environmental)
summary(lmOut1)
```

The summary command produces the following output:

Call: lm(formula = ozone ~ radiation * wind, data = environmental)
Residuals:

Min	1Q	Median	3Q	Max
-48.680	-17.197	-4.374	12.748	78.227

Coefficients:

	Estimate	Std. Error	t value	Pr(>|t|)
(Intercept)	34.48226	17.62465	1.956	0.053015 .
radiation	0.32404	0.08386	3.864	0.000191 ***
wind	-1.59535	1.50814	-1.058	0.292518
radiation:wind	-0.02028	0.00724	-2.801	0.006054 **

Signif. codes: 0 '***' 0.001 '**' 0.01 '*' 0.05 '.' 0.1 ' ' 1
Residual standard error: 24.15 on 107 degrees of freedom
Multiple R-squared: 0.4875, Adjusted R-squared: 0.4732
F-statistic: 33.93 on 3 and 107 DF, p-value: 1.719e-15

Similarly to the factorial ANOVA, it makes sense to examine this output in reverse order. All the way at the bottom we have an F-test on the null hypothesis that R-squared is equal to 0. With the rather large value of $F(3,107) = 33.93$, we can reject that null hypothesis, as the associated p-value is less than 0.05 (and also less than 0.01 and 0.001). The observed R-squared value of 0.4875 may be interpreted as saying that the independent variables, as a set, accounted for about half of the variability in ozone levels. Of course, to be safe, we would want to examine and report the adjusted R-squared: at 0.47 it has been adjusted slightly downward to avoid the capitalization on chance that often occurs when we are creating models from samples of data.

Moving on now to the coefficients, we should interpret the interaction term first, in case it influences how we make sense out of the linear main effects. In this case the radiation:wind coefficient is statistically significantly different from 0 with a t-value of -2.8 and a p-value less than 0.05. This suggests that the interaction we plotted in Figure 9.3 is statistically significant. Looking now at the main effects, we see that wind is not significant, but radiation is significant. We did not have a research hypothesis about wind (other than its interaction with radiation), but the significant coefficient on radiation—together with the shape of the plot in Figure 9.5—supports our prediction that on sunnier days the greater amount of solar radiation would be associated with higher levels of ozone.

Finally, at the beginning of the output, we see that the median of the residuals is about -4.4, suggesting that the distribution of residuals is negatively skewed. This raises the possibility of a nonlinear relationship between the independent variables and the dependent variables. If we were responsible for writing a report on these data, we would want to make additional efforts to assess whether this nonlinearity was problematic for the interpretation of our variables. We can explore this issue more deeply by examining the residuals with respect to different levels of the independent and dependent variables. First, let's look at the residuals versus the independent variable radiation, using plot(environmental$radiation, residuals(lmOut1)), and abline(h = 0). The graph appears in Figure 9.6.

Note that cloud of points in Figure 9.6 is for the most part centered around the horizontal line that goes through 0 on the Y-axis. In an ideal regression, the dispersion of points above and below the line would be random and uniform at all levels of the independent variable. You can see, however, that as we get up to higher levels of radiation, the errors of prediction are much more highly dispersed around that horizontal line. The technical word for this is **heteroscedasticity,** meaning that the variance of the residuals is different at different levels of the independent variable. Heteroscedasticity is the opposite of **homoscedasticity,** which as you might imagine signifies that the variance of residuals is about the same across all levels of the independent variable. Likewise if we examine the residuals versus the dependent variable using plot(environmental$ozone,residuals(lmOut1)) and abline(h = 0), we get Figure 9.7.

This diagram also shows some anomalies. At low levels of ozone, the errors of prediction are mainly negative, while at higher levels of ozone they are mainly positive. Another way to think about this is that we are predicting a little too high at low levels of ozone and a little too low at high levels of ozone. This result suggests nonlinearity in the relationship between radiation and ozone (or possibly in the relation of wind to ozone). We could and should do some additional analyses—for instance, by including a quadratic term in the prediction of ozone from radiation—a squared version of the radiation variable.

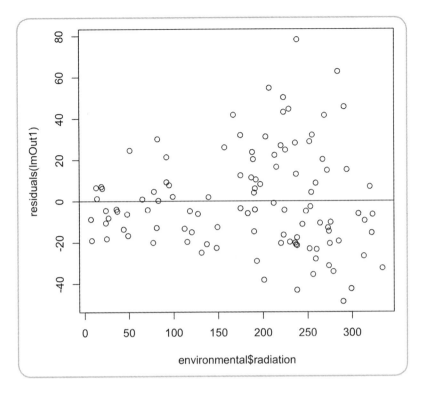

FIGURE 9.6. Regression residuals versus the independent variable radiation.

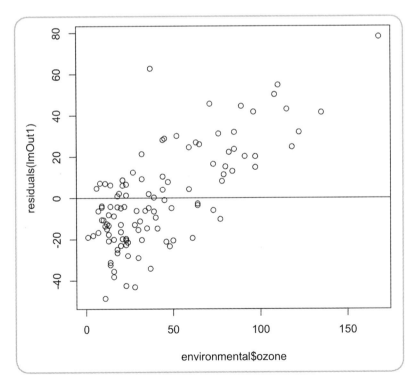

FIGURE 9.7. Regression residuals versus the dependent variable ozone.

Diagnosing Residuals and Trying Alternative Models

In this chapter we conducted a more thorough analysis of residuals than we did earlier in the book. In the case of the environmental data set, our inspection of residuals diagnosed both some heteroscedasticity and the possibility of a nonlinear connection between one of the predictors and the outcome variable (note that the two problems may be related). While I am glad we inspected the residuals and found these anomalies, if we had scrutinized our variables beforehand, we might have noticed that radiation and ozone have a nonlinear relationship between them. You can see this for yourself by creating scatterplots of each pair of variables:

```
pairs(environmental,panel=panel.smooth)
```

The pairs() command creates a square matrix of scatterplots, while the "panel=panel.smooth" argument displays a smooth, fitted curve on each of the scatterplots. To the extent that these fitted curves are indeed curved, rather than reasonably straight, that suggests the possibility of a nonlinear relationship between the two variables depicted. The scatterplot for radiation versus ozone shows a linear relationship at low levels of radiation, but as radiation gets higher, ozone levels are much more dispersed, and the connection between radiation and ozone bends strongly to a new slope. The curved pattern between radiation and ozone is quite likely what caused the anomalies we observed in the residuals from the multiple regression analysis.

In some analytical situations we can address this kind of anomaly either through transforming one of the variables or by adding higher degree terms to the regression. Common transformations include squaring, square rooting, or taking the log of the variable. You can experiment with those kinds of transformations using functions such as sqrt() and log(). Simply add a transformed version of a variable to your data frame and use that in place of the original. For the environmental data set, I had an intuition that we might introduce a squared version of the radiation variable alongside the linear version to see if that would improve our regression model. Including a squared, or quadratic, version of a variable allows lm() to model a possible curvilinear relationship in addition to the basic linear relationship that may exist between a predictor and the outcome variable. The code to accomplish the addition of a quadratic term for radiation looks like this:

```
env <- environmental            # Make a copy of the data set
env$radSqr <- env$radiation^2   # Add a squared version of radiation
# Include both radiation and radSqr in our new lm() model
lmOutQuad <- lm(ozone ~ radiation + wind + radSqr + radiation:wind, data=env)
summary(lmOutQuad)              # Review the model containing the new term
```

(continued)

You should run this code in order to examine the detailed output. I found that the coefficient on the variable "radSqr" was statistically significant and that the total R-squared for the model had risen from 0.499 in the original model to 0.511 in this model with radSqr. I inspected the residuals again, and although there was still some hint of heteroscedasticity and nonlinearity, these were not as pronounced as before. Of further interest, the interaction term was still significant and the linear effect of radiation was strengthened in comparison to the original analysis. These are all hints that the addition of the quadratic term was useful.

To back up these hints, we would want to compare the models with and without the quadratic term, to see if the addition of the quadratic term significantly improved the R-squared. We saw a preview of this idea when we compared nested ANOVA models with the output of anovaBF(). Later in the chapter I will introduce a procedure called modelCompare() that facilitates such a comparison between model outputs generated by lm().

One other important issue that we must consider is *centering*. Statisticians who study regression interactions, often called "moderated multiple regression," have found that the interaction term may have spuriously high correlations with the other independent variables. From a mathematical standpoint the interaction term is literally the product of the independent variables that go into it. If, for example, you take the product of two vectors of positive numbers, the product will tend to correlate quite highly with each of the two multiplicands. You can try it yourself with the following code:

```
set.seed(1)     # Control randomization so you get the same result as me
a <- runif(100, min = 0, max = 1)   # A random uniform variable
b <- runif(100, min = 0, max = 1)   # Another random uniform variable
c <- a * b                          # c is the product of a and b
cor(data.frame(a,b,c))              # Get a correlation matrix of all three
```

Remember that runif() provides a uniform distribution of values between 0 and 1 as a default. If you run this code you will find that the interaction term "c" correlates at a very high level with both "a" and "b" even though a and b are uncorrelated with each other. For a contrasting result, try this:

```
a <- runif(100, min=-1, max=1)     # Center a on zero
b <- runif(100, min=-1, max=1)     # Center b on zero
c <- a * b                         # c is a product of a and b
cor(data.frame(a,b,c))             # Show a correlation matrix for all three
```

By centering a and b around 0, we eliminated most of the correlation between a:c and b:c. As an exercise, create a new variable that is the product of radiation and wind and see how highly it correlates with each of these two variables. You will find that it correlates especially highly with wind, which in turn would impact the results of the regression analysis. The solution to this problem that many statisticians recommend is *centering*. By centering each variable around a mean of 0, we can eliminate the spurious correlations with the interaction term. In R we can use the scale() command to accomplish this in one step:

```
# Rerun the analysis with centered variables
stdenv <- data.frame(scale(environmental,center=TRUE,scale=FALSE))
lmOut2 <- lm(ozone ~ radiation * wind,data=stdenv)
summary(lmOut2)
```

The scale command has two control parameters, center and scale, that default to true. In the command above, I have included center=TRUE to emphasize that this is what scale() is doing. The other control parameter, also known as scale, will normalize each variable to have a standard deviation of 1, when scale=TRUE. So scale(), with the default settings, creates standardized variables (mean = 0, SD = 1) from a data matrix or data frame. I could have completely standardized these variables, instead of just centering them, and gotten functionally equivalent results, but by leaving the standard deviations of the variables alone, you will see from the output that the coefficient on the interaction term is identical to our previous results:

```
Call: lm(formula = ozone ~ radiation * wind, data = stdenv)
Residuals:
    Min      1Q  Median      3Q     Max
-48.680  -17.197  -4.374  12.748  78.227
Coefficients:
```

	Estimate	Std. Error	t value	Pr(>|t|)
(Intercept)	-0.83029	2.31159	-0.359	0.72016
radiation	0.12252	0.02668	4.592	1.20e-05 ***
wind	-5.34243	0.65270	-8.185	6.17e-13 ***
radiation:wind	-0.02028	0.00724	-2.801	0.00605 **

```
---
Signif. codes: 0 '***' 0.001 '**' 0.01 '*' 0.05 '.' 0.1 ' ' 1
Residual standard error: 24.15 on 107 degrees of freedom
Multiple R-squared: 0.4875, Adjusted R-squared: 0.4732
F-statistic: 33.93 on 3 and 107 DF, p-value: 1.719e-15
```

Notice too that the overall R-squared and the F-test on the R have remained identical to our previous results. So centering has not affected the overall quality of our model. What has changed quite substantially, however, are the coefficients and significance tests for our main effects. Notice that both radiation and wind coefficients are now significant at $p < .001$. This is because we have removed the spurious correlations between these predictors and the interaction term by centering the variables. We have not changed the overall amount of variance accounted for in the model but we have partitioned that variance differently in this centered version of our regression model. If you were conducting research using this data set, it would be preferable to report these results instead of the ones we examined earlier in the chapter.

Most contemporary reports of moderated multiple regression—that is, regressions with interaction terms—show results based on centered variables. As a diagnostic, you can always create a product term yourself, add it to your data set, and correlate it with the independent variables to see if it has unusually high correlations with the predictors. If it does, you should then try a centered version of your regression model. In the case of our model, if we had developed a substantive hypothesis about the main effect of wind, the first model we ran (without the centered variables) would have led us astray, as wind has a quite strong (negative) correlation with ozone and as a result would typically have a significant coefficient in our regression model.

To put the icing on our moderated regression cake, we should compare the models with and without the interaction term. The lmSupport package contains a function called modelCompare() which allows calculation of **delta-R-squared.** Delta is the Greek letter that looks like a triangle, and it is often used as an abbreviation for the amount of change in some quantity. So the idea of delta-R-squared is that we are looking for a change in R-squared. The trick, of course, is not just to look for any minor change, but to see if the difference in two R-squared values is large enough to be statistically significant. When making such comparisons, it is generally required that the models be **nested,** in other words that a simpler model is made somewhat more complex by the addition of one or more new terms (predictors). In that situation, we say that the more complex model is nested inside the simpler model. Here's an example where we compare models with and without the interaction term:

```
lmOutSimple <- lm(ozone ~ radiation + wind,data=stdenv)
lmOutInteract <- lm(ozone ~ radiation + wind + radiation:wind,data=stdenv)
install.packages("lmSupport")
library(lmSupport)
modelCompare(lmOutSimple, lmOutInteract)
```

That final command produces the following output on the console:

```
SSE (Compact) = 66995.86
SSE (Augmented) = 62420.33
Delta R-Squared = 0.03756533
Partial Eta-Squared (PRE) = 0.06829571
F(1,107) = 7.843305, p = 0.006054265
```

The first two lines compare the sum of squared errors for each model. The more complex model (containing the interaction) generally contains the lower value for sum of squares because it has more elements that can account for variance in the dependent variable. The third line shows the delta-R-squared: the larger this value is, the stronger the interaction effect. The fourth line is the partial eta-squared of the change in model, expressed as the difference in the sums of squares divided by the sum of squares of the compact model—in effect, the proportion of error reduction accomplished by the more complex model. The final line conducts an F-test on the delta-R-squared. In this case, $F(1,107)$ = 7.84 is statistically significant at $p < .01$, so we reject the null hypothesis of no difference between the R-squared values of the two models. These results provide evidence of an interaction effect between radiation and wind, over and above the linear effects of these two variables on ozone levels.

BAYESIAN ANALYSIS OF REGRESSION INTERACTIONS

As always, we can enhance our analytical evidence by conducting a parallel set of Bayesian analyses. Because we obtained sensible results from our centered variables using the conventional regression technique, let's continue to use the centered data for our Bayes analysis as well. Because we know that we will want to compare models with and without the interaction term, let's run both models right now:

```
lmOutBayes1 <- lmBF(ozone ~ radiation + wind, data=stdenv)
lmOutBayes2 <- lmBF(ozone ~ radiation + wind + radiation:wind, data=stdenv)
```

The first model produces a Bayes factor of 7.098e+11, extremely strong odds in favor of the alternative hypothesis that radiation and wind predict ozone, with nonzero coefficients. Likewise the second model produces a Bayes factor of 4.27e+12, also a very strong indication in favor of the alternative hypothesis. You can see from the code that the only difference between the first model and the second model is that the second model includes the interaction term. We can compare those models and obtain a Bayes factor for the result simply by creating a fraction:

```
lmOutBayes2/lmOutBayes1
```

This model comparison produces the following output:

```
Bayes factor analysis
--------------
[1] radiation + wind + radiation:wind : 6.019568 ±0.01%
```

Against denominator:
ozone ~ radiation + wind

Bayes factor type: BFlinearModel, JZS

The results show the odds of 6:1 in favor of the model that includes the interaction, a good solid result worth reporting. One might summarize these results by saying that the Bayes factors show that the odds are in favor of a model that includes both main effects of radiation and wind as well as the interaction between them. The interpretation of these effects goes all the way back to the beginning of the section where we examined the interaction plot. A substantive interpretation of these data should include such a plot as well as an explanation of the difference in slopes between the two regression lines. To the extent that there are theoretical explanations for the observed data, these should be included in the explanation of the statistical output.

As always, we can generate a posterior distribution to obtain a much more detailed view of each of the coefficients, as well as of the overall R-squared. Let's repeat the run of our second model, this time with posterior=TRUE and iterations=10000:

```
mcmcOut <- lmBF(ozone ~ radiation + wind + radiation:wind, data=stdenv,
    posterior=TRUE, iterations=10000)
summary(mcmcOut)
```

The summary() command produces the following output:

1. Empirical mean and standard deviation for each variable,
   plus standard error of the mean:

	Mean	SD	Naive SE	Time-series SE
radiation	0.11865	0.02658	0.0002658	2.697e-04
wind	-5.15413	0.65591	0.0065591	7.022e-03
radiation:wind	-0.01962	0.00717	0.0000717	7.291e-05
sig2	591.31645	81.04671	0.8104671	8.464e-01

2. Quantiles for each variable:

	2.5%	25%	50%	75%	97.5%
radiation	0.06672	0.10064	0.11885	0.1364	0.170465
wind	-6.44187	-5.58884	-5.15419	-4.7171	-3.854739
radiation:wind	-0.03380	-0.02438	-0.01966	-0.0148	-0.005604
sig2	455.00786	534.59736	583.59619	640.1165	770.333051

Because we have had a lot of practice in interpreting these tables, I will not bother producing histograms of the posterior distributions, although you could do so yourself with commands like this: hist(mcmcOut[,"wind"]). The mean values of the distributions match the coefficients shown in the conventional regression output fairly well, although in most cases they are slightly smaller.

The 95% HDIs for each coefficient show us the likely distribution of the population values for each coefficient. Looking closely at these, the 95% HDI for radiation does not straddle 0. Likewise, the HDI for wind and for the interaction term also do not straddle 0. These indications firmly back up the results of the conventional analysis by showing that the coefficient of radiation is credibly positive in the population, somewhere between 0.067 and 0.170. Likewise the coefficients on wind and the interaction term are both credibly negative in the population.

You may remember the computation we used in Chapter 8 to calculate the distribution of R-squared values from the posterior distribution of sig2. Here it is again for your convenience:

```
rsqList <- 1 - (mcmcOut[,"sig2"] / var(stdenv$ozone))
mean(rsqList)              # Overall mean R-squared
quantile(rsqList,c(0.025))
quantile(rsqList,c(0.975))
```

Output from the mean() command and the two calls to quantile() reveal that the mean of the posterior distribution of R-squared is 0.47, just slightly underneath the adjusted R-squared of the conventional model generated by lm(). The 95% HDI for R-squared ranges from a low of 0.30 up to 0.59, giving us a view of the most likely values of R-squared in the population. These results concur with the findings of the conventional analysis, but also give us a clearer view of the uncertainty surrounding our estimates of the coefficients and R-squared.

Let's close our analysis by calculating and plotting some predictions for our moderated regression example. I'm going to double back to the high- and low-wind data sets we examined earlier in the chapter. To ensure that my plots have a realistic Y-intercept, I am also going to go back to the original moderated regression analysis we did before centering the variables. The lmSupport package provides a helpful command called modelPredictions() that allows us to generate predicted values of our outcome variable using the model coefficients in the output from an lm() command:

```
# Set wind to a constant
loWind$wind <- mean(loWind$wind) - sd(loWind$wind)
hiWind$wind <- mean(hiWind$wind) + sd(hiWind$wind) # Here as well
loWindOzone <- modelPredictions(lmOut1, Data=loWind, Type = 'response')
hiWindOzone <- modelPredictions(lmOut1, Data=hiWind, Type = 'response')
plot(loWind$radiation,loWindOzone$Predicted,xlim=c(0,350),ylim=c(10,90))
points(hiWind$radiation,hiWindOzone$Predicted,pch=3)
```

The results of the plot() and points() commands appear in Figure 9.8. Note that the code above uses an old statistician's trick to get two clean, straight lines in the plot. Rather than using real values of the wind variable, we set wind to a constant. For the low-wind condition, we choose a value one standard deviation below the mean. For the high-wind condition we use a value one standard deviation above the mean. So we are taking a realistic set of values for radiation (the original raw data actually) and calculating the predicted values of ozone with the wind variable set to a constant. In the first plot, represented in Figure 9.8 by little circles, we see a prediction line for low-wind conditions. This line is quite steep, showing a close connection between radiation and ozone on low-wind days. In the second plot, represented in Figure 9.8 by little plus signs, we see a prediction line for high-wind conditions. This line is quite shallow, showing minimal connection between radiation and ozone on high-wind days. Although it is uncommon to see this kind of plot in a journal article or other research report, researchers often find it helpful to visualize lines like this to help them properly describe the configuration of the regression interaction.

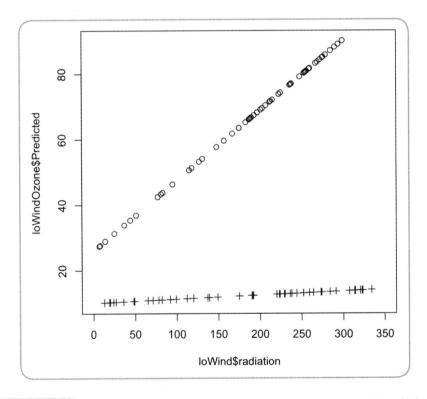

FIGURE 9.8. Contrasting regression lines for low wind versus high wind.

CONCLUSION

Interactive effects, or interactions, as they are commonly known among researchers, open up a variety of additional opportunities for hypothesis testing by letting us examine how independent variables may work together to create additional effects on a dependent variable. We looked at two examples of this. In the first example, we wondered whether a high-protein diet would be more effective at causing weight gain in rats if that protein came from meat instead of cereal. In the second example, we wondered whether the relationship between solar radiation and the ground concentration of ozone would be stronger on days with less wind and weaker on days with more wind. Obviously, if you want to examine an interaction you have to have at least two predictor/independent variables. The interaction term is quite literally the product of those two predictor variables. Note that higher order interactions—for example, the product of three predictor variables—are also possible but we did not examine them here.

In addition to examining interactions, we took a look at a regression diagnostic by plotting the residuals against both an independent variable and the dependent variable. For ordinary least squares modeling, which we were doing throughout this chapter, residuals should have a mean of 0 and be normally distributed. Residuals should also have about the same level of variance regardless of where we are along the range of the other variables—that is the concept of homoscedasticity. When we examine residuals against the dependent variable, we may also uncover evidence of nonlinear relationships between independent and dependent variables. There are ways of dealing with this issue that we touched upon in "Diagnosing Residuals and Trying Alternative Models."

We also examined a commonly used technique called centering that helps us to ensure that there are not spurious correlations between the interaction term and other variables in a moderated multiple regression analysis. Centering is very commonly used by researchers when testing for interactions in linear regression.

Finally, as has been the case throughout the book we compared the standard methods of the null hypothesis significance test (NHST) to the more contemporary Bayesian methods of directly exploring the alternative hypotheses. The NHST gives us a go/no–go decision-making technique that is easy to use (though not always easy to understand) and also easy to abuse. The Bayes factor strategy, together with the examination of posterior distributions, allows us to examine the strength of evidence in favor of the various alternative hypotheses. We must, of course, make our own informed judgments about the practical impact of our statistical findings.

EXERCISES

1. The data sets package in R contains a data frame called "C02." Reveal the variable names and data types in the data set using appropriate R commands and identify the names and levels of all the independent variables that are factors. What are the possible two-way interactions that could be explored in this data set (i.e., name the combinations of factors)? What are the possible dependent variables that one could examine in an ANOVA analysis?

2. Use the interaction.plot() command in the HSAUR package to display a means plot of the "uptake" variable, using "Type" and "Treatment" as the factors. Interpret the results: Without knowing any of the statistics, would you guess that there is or is not an interaction between Type and Treatment?

3. Use aov() to produce significance tests of the main effects of Type and Treatment and the interaction between Type and Treatment using uptake as the dependent variable. Make sure to state each null hypothesis and then use the correct language to describe the outcomes of the significance tests. Make sure to describe how the degrees of freedom work for this model. Make sure to report "omnibus" statistics as well (e.g., the overall R-squared).

4. Use anovaBF() to examine the main effects and interaction as described in Exercise 3. Interpret the results in your own words. If you also did Exercise 3, contrast the results from the traditional ANOVA analysis with the results of the Bayes Factor ANOVA. Important note: The anovaBF() function is picky about how the data sets it analyzes are stored. Use newCO2 <- data.frame(CO2) to convert the CO2 data set to a data frame. Then conduct your analysis on newCO2. Make sure to use the output object from the anovaBF() procedure to create an odds ratio that compares the complete model (with the interaction) against a main effects-only model.

5. The USJudgeRatings data sets package in R contains a set of 43 ratings of the competence of state judges in the U.S. Superior Court (from the 1970s). Reveal the variable names and data types in the data set using appropriate R commands. In this exercise, we are particularly interested in two variables that we will use as independent variables, FAMI and PREP, as well as one variable that we will use as the dependent variable: RTEN. Use the help command "?USJudgeRatings" to reveal information about the meaning of these three variables and summarize what you find in your own words.

6. Using similar code as that which appeared in the regression interactions section of this chapter, plot regression lines showing PREP (on the X-axis) and RTEN (on the Y-axis). Plot two regression lines, one for the half of the data set where FAMI is below its median value and one where FAMI is at or above its median value. Without knowing any of the statistics, do you think it likely that there is an interaction between FAMI and PREP? Explain why or why not in your own words.

7. Copy USJudgeRatings into a new data frame object and center the variables. Use an appropriate R function to report the means of the variables and confirm that they are all 0. Report the means and say in your own words why centering is important when testing regression interactions.

8. Conduct a regression analysis of the linear and interactive effects of PREP and FAMI on RTEN using the lm() function. Interpret the results in your own words, making sure to report the outcomes of the significance tests.

9. Repeat Exercise 8 using lmBF() to conduct one regression analysis that includes the interaction term and one that does not. Interpret the results. Make sure you report an odds ratio for the comparison of a model with linear effects versus a model with linear and interaction effects. Interpret the results in your own words.

10. Bayesian bonus question: Obtain posterior distributions for the regression coefficients from Exercise 9. Also use appropriate calculations to obtain a posterior distribution for R-squared. Report and interpret your results.

11. Graphical bonus question: Redo Exercise 8 without centering the variables. Use the results, in combination with the modelPredictions() function, to create prediction lines for PREP versus RTEN, with FAMI held constant at two different values—one standard deviation above the mean and one standard deviation below the mean.

## CHAPTER 10

Logistic Regression

In the previous three chapters we explored the general linear model by focusing on the prediction or modeling of **metric** variables. I use the catch-all term "metric" to refer to a variable with an ordered set of values that could be placed on a number line or the axis of a graph. If you would like to explore the distinctions among different kinds of variables more deeply, try searching online for "nominal, ordinal, interval, and ratio." The general linear model includes both ANOVA and linear multiple regression. When we applied the ANOVA analysis, we used categorical independent variables (AKA "factors") to predict one metric outcome variable. Similarly, we used linear multiple regression to predict a metric outcome variable from one or more metric predictor variables. Although we did not explore this idea in depth, both ANOVA and linear multiple regression actually allow for a mix of categorical and metric independent variables. But with the general linear model the dependent variable is always metric data.

In this chapter we expand our horizons by examining "logistic" regression, which provides the opportunity to create prediction equations where the dependent variable is **categorical**—in the simplest case a dependent variable that can take either of two states (e.g, on/off, up/down, true/false). So while in previous chapters we considered techniques that fit the relatively simple "general linear model," in this chapter we examine the "*generalized* linear model." I know it seems a bit goofy to have two names that are so similar to one another, but there are some meaningful similarities. In both cases we are predicting a dependent variable from a set of one or more independent variables. In both cases we can have interaction terms that look at the combined effects of two or more independent variables. In both cases our independent variables can be a combination of categorical and metric predictors.

The major difference is that the *generalized* linear model uses a so-called link function that can take many shapes and not just a straight line (as is the case with the *general* linear model). Whereas we used the **least squares** method to fit our ANOVA and regression models, these new techniques create a model through an iterative series of approximations—often using a strategy called **maximum likelihood estimation.** At first, that may seem similar to some of the Bayesian techniques we have explored, but there are important distinctions. For instance, with maximum likelihood estimation, there is just one point estimate for each population parameter and as a result there is no posterior distribution.

Back to the idea of a link function: for logistic regression, we will use a link function called the **inverse logit.** The inverse logit function will allow us to create models that predict binary outcomes. You can think of the inverse logit function as a series of probabilities related to whether a binary outcome is true or false. The inverse logit function uses the natural logarithm to develop a beautiful S-shaped curve like this:

```
# Create a sequence of 100 numbers, ranging
# from -6 to 6 to serve as the X variable
logistX <- seq(from=-6, to=6, length.out=100)
# Compute the logit function using exp(), the inverse of log()
logistY <- exp(logistX)/(exp(logistX)+1)
# Now review the beautiful S curve
plot(logistX,logistY)
```

The results of the plot() command appear in Figure 10.1. The inverse logit function is often called the **logistic curve** by mathematicians. Note that the exp() function that is built into R simply raises Euler's number "e" (a transcendental number approximately equal to 2.718) to the power of the value provided in the argument. Leonhard Euler (1707-1783) was a Swiss mathematician who made huge contributions to the foundations of modern statistics. Interesting statistical trivia: the standard normal distribution is also a function of powers of Euler's number.

Here's an example to help you think about the logistic curve. Let's imagine that the variable on the *X*-axis is an index of palm sweat, or what spies call the galvanic skin response (GSR). GSR sensors use a small electric current to measure the resistance across the surface of the skin. Some research shows that when an individual is telling a lie, their palms sweat slightly more and this can be detected with a GSR device. So let's think of −6 on our *X*-axis in Figure 10.1 as totally not sweaty, completely dry-like-a-desert palms. In contrast, with a measure of +6 the sweat is simply dripping off the person's palms. Now you can read the *Y*-axis as the probability that the person is telling a lie. So when GSR is between −6 and about −2, the chances that the person is telling a lie are less than 20%. In contrast, GSR readings between +2 and +6 show that the chance that the person is lying is 80% or more.

You will note that there is a steep transitional area between −2 and plus 2 where the probabilities shift very quickly from truth-more-likely to

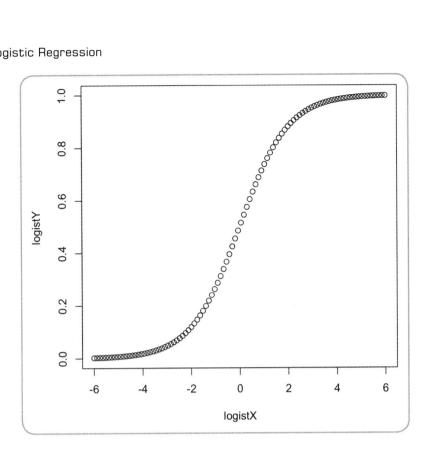

FIGURE 10.1. An example of the logistic curve.

lie-more-likely. Even though everything is in transition here, keep in mind that in our example we are trying to model the reality that either the person is lying or the person is not lying. Therefore, it is that binary outcome (truth or lie) that is our dependent variable, even if we choose to model it as a continuous logit curve. In fact, this is the cool trick that logistic regression accomplishes for us: while it is difficult to statistically model a step function—that is, something that transitions instantaneously from 0 to 1, it is much easier to model a continuous function, such as the logit. So we use the logit to stand in for the binary response and we find out what we need by looking at probabilities (in the form of odds or log–odds). Thus, the logit is our *link function* that we use in this application of the generalized linear model.

Let's create some fake variables so that we can explore these ideas a little further. First, let's create a more realistic predictor variable. As has been true in the rest of the book, we commonly work with independent variables/predictors that are metric and approximately normally distributed. So let's put in a random normal variable in place of the X that we created before:

```
# Create a random, standard-normal predictor variable
set.seed(123)
logistX <- rnorm(n=100,mean=0,sd=1)
```

I use set.seed() so that you will have the same random numbers and can create the exact same results as are shown below. Our X variable now contains a random normal variable with mean of 0 and standard deviation of 1. Next, let's re-create our logit curve based on our new values of X. Remember that exp() is the inverse of the natural log:

```
# Create an outcome variable as a logit function of the predictor
logistY <- exp(logistX)/(exp(logistX)+1)
```

Now that we have created our Y variable, let's transform it into the binary outcome that we are actually trying to model. Keep in mind that for the purposes of this demonstration we are creating this fake example in reverse of what really happens inside a logistical regression analysis: in a real analysis you start with the binary outcome variable and it is the logistic regression analysis process that converts the Y variable into the logistic function for you. For this fake example, however, we will synthesize our binary Y variable using the round() function to split the responses at the inflection point of the probability curve (0.5). This works nicely because the values of logistY vary from 0 to 1, and the round function makes everything less than 0.5 into 0 and everything at 0.5 and above into 1:

```
# Make the dichotomous/binomial version of the outcome variable
binomY <- round(logistY)
```

Finally, to make our example more realistic, I will add some noise to the predictor variable so that the relationship between the X and the Y is imperfect. After all, if I could use a GSR detector to perfectly predict lies and truth, I would be selling GSR machines instead of writing a data analysis textbook:

```
# Add noise to the predictor so that it does not perfectly predict the outcome
logistX <- logistX/1.41 + rnorm(n=100,mean=0,sd=1)/1.41
```

Now let's review the new variables we have created with some visualizations. The first visualization, a regular scatterplot, appears in Figure 10.2.

```
plot(logistX, binomY)
```

I think you can see that the plot in Figure 10.2 is difficult to interpret. The beautiful S-shape is gone: it is almost as if a giant magnet pulled all of the low values to the bottom of the plot and the high values to the top of the plot. In fact, if you think about how the round() function split up all the values of X to make binomY either valued 0 or 1, that is essentially what happened. Unfortunately this resulted in a scatterplot that is very hard to interpret. You should be able to see that for quite a lot of the low values of X, the Y value is now set to 0. Likewise, for many of the high values of X, the Y value is set to 1. We can make

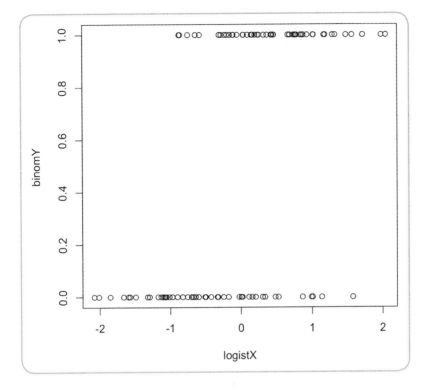

FIGURE 10.2. Traditional scatterplot of continuous/metric *X* variable and dichotomous or binomial *Y* variable.

a better plot just by temporarily reversing our view of *Y* and *X*. In the code below we coerce our dichotomous *Y* variable into a factor with some friendly labels, put everything in a data frame to get ready for our logistic regression analysis, and then use a box plot to visualize the result:

```
binomY <- factor(round(logistY), labels=c('Truth','Lie'))
logistDF <- data.frame(logistX, logistY, binomY) # Make data frame
boxplot(formula=logistX ~ binomY, data=logistDF, ylab="GSR", main="Lie
    Detection with GSR")
```

The diagram in Figure 10.3 is much easier to interpret, even though we now have the outcome variable on the *X*-axis and the predictor on the *Y*-axis—backwards from how we normally think about this kind of thing (e.g., if we were doing an ANOVA, truth/lie would be the predictor and GSR the outcome, but of course that makes no sense here). We have the distribution of our *X* variable shown on the vertical axis, with a separate boxplot of *X* across the two different categories of the *Y* variable. I have labeled the diagram thoroughly so that you can see what is going on. For relatively low values of GSR, in other words, those cases where the subjects' palms are pretty dry, you can

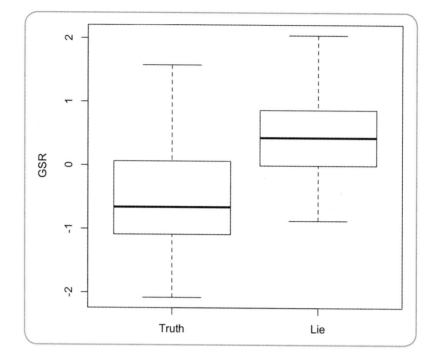

FIGURE 10.3. Box plot with dichotomous *Y* variable displayed as a grouping variable.

see that most of these fall into the "Truth" category for the dependent variable. Likewise, for the relatively high values of GSR, that is, the cases where the subjects' palms are pretty sweaty, you can see that most of these fall into the "Lie" category for the dependent variable.

Of course there is also some degree of overlap: there are a few cases of high GSR (sweaty palms) in Figure 10.3 that are nonetheless designated as truth and a few cases of low GSR (dry palms) that are designated as lie. Some people have naturally dry or sweaty palms, such that GSR is an imperfect predictor of truth or lies. How good a predictor is GSR in our fake example? Now we are ready to run our logistic analysis so that we can answer that question:

```
glmOut <- glm(binomY ~ logistX, data=logistDF, family=binomial())
summary(glmOut)
```

That call to the summary() command yields the following output:

```
Call:
glm(formula = binomY ~ logistX, family = binomial(), data = logistDF)
Deviance Residuals:
    Min      1Q  Median      3Q     Max
-2.3216  -0.7982  0.3050  0.8616  1.7414
```

Coefficients:

	Estimate	Std. Error	z value	Pr(>|z|)	
(Intercept)	0.1199	0.2389	0.502	0.616	
logistX	1.5892	0.3403	4.671	3e-06	***

Signif. codes: 0 '***' 0.001 '**' 0.01 '*' 0.05 '.' 0.1 ' ' 1

(Dispersion parameter for binomial family taken to be 1)
Null deviance: 138.47 on 99 degrees of freedom
Residual deviance: 105.19 on 98 degrees of freedom
AIC: 109.19
Number of Fisher Scoring iterations: 4

Starting at the top, the call to glm() is similar to the call to lm() with one exception: we have specified the "link function"—in this case by indicating family=binomial(). By specifying "binomial()" we invoke the inverse logit or logistic function as the basis of fitting the X variable(s) to the Y variable. R dutifully repeats back the model specification we provided. Next the "Deviance Residuals" (oddly appropriate when discussing lie detection) show diagnostic information about the distribution of the residuals after the model is fit. The mean of the residuals is always 0 (within the limits of rounding error) and you can verify this for yourself with the command mean(residuals(glmOut)). The fact that the *median* residual is slightly positive suggests that the distribution of residuals is slightly negatively skewed. You can verify this with hist(residuals(glmOut)). Remember that the residuals represent errors of prediction, so if there is a residual (or possibly more than one) that is strongly negative or strongly positive, it may suggest a problem such as the presence of an outlier. If you look back at Figure 10.3, however, you will see that there are no apparent outliers, so the slight skewness in residuals is not a problem here.

Next we get to the fun part, the coefficients. The Y-intercept is not significantly different from 0, an expected result when our X variable was centered on 0. The coefficient on logistX shows the strength of prediction of our Y variable (truth or lie) based on our X variable (GSR). We immediately observe that the coefficient is significantly different from 0, as this is supported by the "Wald" z-test (conceptually similar to a t-test) and the associated p-value. So we reject the null hypothesis that the coefficient on logistX is 0 in the population. What is the nature of this coefficient, however, when we are predicting a binary variable?

It was easy to imagine what the coefficient meant for linear regression—it was simply the slope of a line. It's trickier, however, to think of the slope of a line in relation to a logistic curve that has different slopes at different points on the X-axis. Instead, what we have here is something fairly complex but worth taking the time to grasp: it is the **logarithm of the odds of the Y variable,** commonly referred to by statisticians as **log-odds.** So the log-odds coefficient on our predictor is 1.5892, but what does that really signify? Because most people are not used to thinking in terms of log-odds, we can transform this coefficient so that it can be interpreted as regular odds instead of log-odds:

exp(coef(glmOut)) # Convert log odds to odds

That provides the following simple output:

```
(Intercept)    logistX
1.127432       4.900041
```

This is much better! Now when we look at the transformed coefficient for logistX we can see that a one unit change in logistX gives us a 4.9:1 change in the odds of binomY. So for example, if the GSR variable moved from a value of 0 to a value of 1, it is almost five times more likely that the subject is lying. You can make the statement in the opposite direction as well. If the value of the GSR variable started at −1 (sort of dry palms) to −2 (really dry palms), it is about five times more likely that the person is telling the truth. So our original log-odds coefficient of 1.5892, which is statistically significantly different from 0, when converted to odds is 4.9 which indicates a 4.9:1 change in odds for each unit change in X.

Of course we know that whenever we are dealing with sample data, the point estimate that comes out of the model-fitting process is a single estimate that may or may not be close to the population value. So we should also look at the confidence interval around the estimate. We can convert the confidence interval directly to odds to increase interpretability:

exp(confint(glmOut)) # Look at confidence intervals around log-odds

That provides the following output:

```
Waiting for profiling to be done ...
               2.5 %        97.5 %
(Intercept)    0.7057482    1.812335
logistX        2.6547544    10.190530
```

We know that the intercept was not significant and this result confirms that because the range of odds for the intercept straddles 1:1 (which is effectively the null hypothesis). We can see that the 95% confidence interval for our logistX variable—representing our GSR measurement—runs from a low of 2.65:1 up to a high of 10.19:1. If you recall your definition of the confidence interval it is that if you constructed a very large number of similar experiments based on new samples, that 95% of the confidence intervals you would calculate would contain the population value. For this one particular confidence interval, we don't know if the population value of the odds ratio actually falls within this confidence band or not, but we can tell from the width of the confidence band that there is a fair amount of uncertainty around our point estimate of 4.9:1.

To close out our consideration of the output of glm() I here reproduce the last few lines from the output shown earlier:

```
Null deviance: 138.47 on 99 degrees of freedom
Residual deviance: 105.19 on 98 degrees of freedom
AIC: 109.19
Number of Fisher Scoring iterations: 4
```

The last line is interesting—it shows how many iterations of model fitting it took to get the final model—but I want to focus on the first three lines. The "Null deviance" shows the amount of error in the model if we pretended, for a moment, that there was absolutely no connection between the X variable and the Y variable. The null model thus represents a kind of worst-case scenario that shows what would happen if our predictor(s) had no predictive value. As such it is a baseline to which we can compare other models. The null model shows 99 degrees of freedom because we started with 100 cases and we lose 1 degree of freedom simply for calculating the proportion of truth to lies in our Y variable (equivalent to the grand mean in ANOVA or linear regression).

The next line of output shows how much the error in the null model is reduced by introducing our X variable. Of course, we lose one degree of freedom by introducing our metric X variable, but on the other hand we also substantially reduce the amount of error in the model by letting the X predict the Y. You can think about it as a comparison between the "Null deviance" line and the "Residual deviance" line. By introducing our predictor into the model (which cost one degree of freedom) we reduced error from 138.47 (null model) down to 105.19 (one predictor model). The difference between the null deviance and the model deviance, in this case (138.47–105.19) = 33.28, is distributed as chi-square and can be directly tested from the output model with this somewhat obscure command:

```
anova(glmOut, test="Chisq") # Compare null model to one predictor model
```

```
Analysis of Deviance Table
Model: binomial, link: logit
Response: binomY
```

Terms added sequentially (first to last)

	Df	Deviance	Resid. Df	Resid. Dev	Pr(>Chi)
NULL			99	138.47	
logistX	1	33.279	98	105.19	7.984e-09 ***

Signif. codes: 0 '***' 0.001 '**' 0.01 '*' 0.05 '.' 0.1 ' ' 1

This is the **omnibus** test for this analysis, equivalent to the significance test of R-squared on a linear regression model (*omnibus* is the Latin word for "all"). You can see the probability of observing a chi-square value of 33.279 on one degree of freedom is extremely low and well below our conventional thresholds of alpha, so we can reject the null hypothesis that introducing logistX into the model caused zero reduction of model error. The omnibus test is statistically significant at $p < .001$. We can consider this rejection of the null hypothesis as evidence that the one-predictor model is preferred over the null model.

One last note about the overall results: the second-to-last line in the output read AIC: 109.19. AIC stands for the Akaike information criterion, originally developed by statistician Hirotugu Akaike (1974). The AIC examines error reduction accomplished by a model in light of the number of parameters (i.e., predictors and their coefficients) that it took to achieve that amount of reduction in error. AIC is good for comparing nonnested models: Let's say we ran one GLM model where we predicted Y using variables A, B, C, D, E, and F. Then in a second run of GLM we predicted Y with D, E, F, and G. These two models are nonnested because of the presence of G in the second model, but we could choose the better model of the two by choosing the model with the lowest AIC. AIC also takes into account the fact that model one had six predictors, while model two only had four. It was easier for model one to achieve more error reduction, but at the cost of more complexity. For our current example, we have only run one model with one predictor, so we have no alternative model to which we can compare our AIC of 109.19.

Let's finish our fake example with a look at the quality of our predicted values of Y (truth vs. lies), based on our observed values of X (GSR). The pair of boxplots in Figure 10.4 compares our model outputs (on the left) with our original data (on the right).

```
par(mfrow=c(1,2)) # par() configures the plot area
plot(binomY, predict(glmOut),ylim=c(-4,4)) # Predicted values
plot(binomY, logistX,ylim=c(-4,4))      # Original data
```

The boxplots in Figure 10.4 show that our model is doing a pretty good job. Looking at the left-hand figure, the X values for GSR are notably lower for truth tellers than they are for liars, just as we would hope. Note, however, that the dispersion of X values appearing in the left-hand boxplot (predictions) is noticeably wider than the X values shown in the right-hand boxplot (actual data). This is a natural effect of error in the model: some of the X values, when run through the model, are predicting the wrong thing—for example, a truth when it really was a lie.

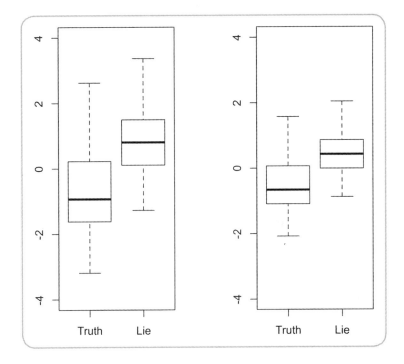

FIGURE 10.4. Comparison of model predictions (left) with original data (right).

A LOGISTIC REGRESSION MODEL WITH REAL DATA

Now let's apply what we have learned to an example with real data. We will use a data set from the "car" package that is called "Chile." The Chile data set contains a poll of Chilean citizens—about 2,700 in all—and their views of the 1988 plebiscite on then-president Augusto Pinochet. We will use two metric variables, age and income, to predict whether the polled individual would vote "Y," in favor of keeping Pinochet in office or "N," meaning opposed to keeping him in office. Because there were two additional options in the poll (abstain and undecided), we must first extract the Y and N responses. I'll also take this opportunity to remove cases that have missing values:

```
ChileY <- Chile[Chile$vote == "Y",]        # Grab the Yes votes
ChileN <- Chile[Chile$vote == "N",]        # Grab the No votes
ChileYN <- rbind(ChileY,ChileN)            # Make a new dataset with those
ChileYN <- ChileYN[complete.cases(ChileYN),]    # Get rid of missing data
ChileYN$vote <- factor(ChileYN$vote,levels=c("N","Y"))  # Simplify the factor
```

As always, there are many ways to accomplish these steps in R, but I built this sequence of code so it is easy to read and understand. In the first step we use a logical condition, Chile$vote == "Y", to select only those rows with a

Multinomial Logistic Regression

In this chapter we examined the simplest form of logistic regression, known as binomial logistic regression. Binomial (meaning two names) means that the outcome variable is binary, also known as dichotomous. While there are lots of outcome variables that take this form—votes (candidate A or candidate B), jury verdicts (innocent or guilty), medical diagnosis results (diseased or not-diseased)—it is quite common to find categorical variables that have more than two categories. In fact, the example used in this chapter of the Chilean plebiscite actually contains four different categories of outcome, Yes, No, Abstain, and Undecided. To demonstrate binomial logistic regression, we blithely eliminated the cases that had Abstain and Undecided as their outcome (this was a relatively small proportion of the total number of cases).

If we want to model a categorical outcome with more than two categories, we need a more flexible tool than binomial logistic regression: a family of analytical techniques known as **multinomial logistic models**. Naturally, because we are using R, there are numerous options for packages that support such analyses. Here is just a small subset of the available packages:

- mlogit—estimates multinomial logistic models using maximum likelihood estimation
- nnet—estimates multinomial logistic models using neural networks
- MCMCpack—the same package we used in this chapter offers a function called MCMCmnl() to estimate multinomial models
- BayesLogit—like MCMCpack(), generates posterior distributions for both binomial and multinomial logistic models.

In binomial logistic regression we used the inverse logit curve to model the probability transition from one state of the outcome variable to the other state. Because there were just the two possibilities for the outcome variable, we could learn everything we wanted by juxtaposing the two states. With multinomial logistic regression things get a little more complex, because we have more than two states of the outcome variable to work with. One way to address this complexity is to designate one of the states of the outcome variable as the baseline and then compare the other states to that baseline. For example, let's say that we were trying to classify colleges based on three characteristics: average class size, yearly tuition, and total students enrolled. Our college-type outcome variable is multinomial and can take on one of three values: community college, liberal arts college, or comprehensive college. In this case, we could designate community college as the baseline. Then, the coefficients for the analysis are configured to model the comparisons between community versus liberal arts, on the one hand, and community versus comprehensive, on the other hand. In fact, one way to think about a multinomial model is that

(continued)

it consists of *k*–1 binomial models, where *k* is the number of categories in the multinomial outcome.

There is one simplification that some analytical procedures allow when the multinomial outcome variable is ordered rather than unordered. For example, think about a situation where your outcome variable pertained to highest educational attainment and it had four categories: primary school-only, secondary school, college, and graduate school. As you have probably surmised, these categories are ordered from least to most educational attainment. We could now create a model where the coefficients on predictors described the log-odds of a transition between any two neighboring categories.

If you are curious to learn more, there is a nice, accessible white paper by Starkweather and Moske (2011) that is available online. Also try searching for "multinomial logistic regression CRAN" and you will find helpful information under the first few links.

Yes vote and place them in a new data frame object called ChileY. Similarly, in the second step we use a logical condition, Chile$vote == "N", to select only those rows with a No vote and place them in a new data frame object called ChileN. Then we use rbind(), which stands for "row bind," to bind the two data sets into a new, combined data frame called ChileYN. In the next step, I use the complete.cases() command to select only those rows that contain values for each of the variables in the data set. Note that complete.cases() spits out a list of row numbers. Finally, I redefine the "vote" variable to ensure that it is a factor with just two levels, "N" and "Y." You can use dim(ChileYN) to verify that the resulting data set contains 1,703 rows of data. You can also use table(ChileYN$vote) to get a tally of the Yes and No votes. As a point of trivia, No votes represent 51% of the cases in this data set, whereas in the actual plebiscite back in 1988, the No side won with 56%.

You may want to review box plots that represent ranges for each of the predictors, divided up by Yes and No votes. With our data set all nicely organized into a data frame, that's easy for us to do with the following code:

```
par(mfrow=c(1,2))
boxplot(age ~ vote, data=ChileYN)
boxplot(income ~ vote, data=ChileYN)
```

The plot on the left-hand side of Figure 10.5 suggests that younger voters may be more likely to vote no. The plot on the right-hand side suggests that wealthier voters might be more likely to vote no. For both predictors, there is substantial overlap in the distributions of the predictors for the Yes and No votes so it is hard to say whether or not these differences are simply due to sampling error. Let's now use logistic regression to see if we can significantly

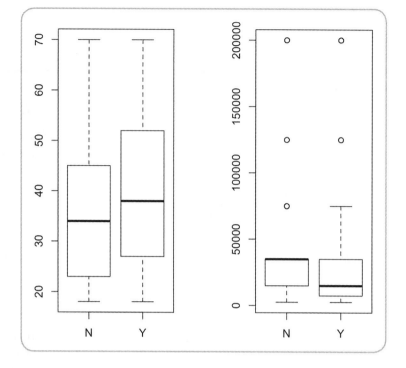

FIGURE 10.5. Predicted values versus residuals for logistic regression of Seatbelt data.

predict a Yes or No vote based on the age and income level of a person who responded to the poll:

```
chOut <- glm(formula = vote ~ age + income, family = binomial(),
    data = ChileYN)
summary(chOut)
```

The key elements of the output from the summary() command appear below:

```
Coefficients:
            Estimate Std.       Error     z value      Pr(>|z|)
(Intercept)  -7.581e-01    1.418e-01     -5.346     9.01e-08 ***
age           1.924e-02    3.324e-03      5.788     7.11e-09 ***
income       -2.846e-07    1.142e-06     -0.249     0.803
Null deviance: 2360.3 on 1702 degrees of freedom
Residual deviance: 2326.0 on 1700 degrees of freedom
AIC: 2332
```

The output shows that intercept is significantly different from 0. In a logistic regression like this, the intercept represents the log-odds of a "Yes" vote when both age and income are equal to 0. Since we can assume that the research design did not include polling newborns with no money, the value of the intercept is not very meaningful to us, but we must keep it in the model to make sure that the other coefficients are calibrated correctly.

The coefficient on the age predictor is statistically significant, based on the Wald's z-test value of 5.79 and the associated p-value. Because the scientific notation $p = 7.11e{-}09$ means that $p < .001$, we can reject the null hypothesis that the log-odds of age is 0 in the population. The Wald's z-test is calculated by dividing the coefficient value by the standard error (labeled Std. Err. in the output). As always, plus or minus 2 standard errors also provides an indication of the width of the confidence interval around the coefficient estimate.

The very tiny coefficient on the income variable is not significantly different from 0, based on a Wald's z-test value of -0.249 and an associated p-value of 0.803. Thus we *fail to reject* the null hypothesis that the log-odds of income is equal to 0 in the population. Keeping in mind that all of these coefficients are log-odds values, let's use exp(coef(chOut)) to convert them into regular odds (some analysts call them "straight" odds):

```
(Intercept)    age         income
0.4685354      1.0194293   0.9999997
```

The intercept represents odds of 0.46:1 for a Yes vote by a penniless newborn. The odds of 1.019:1 for age show that for every additional year of age, a person is about 1.9% more likely to vote Yes. In the case of income, the odds are almost exactly 1:1, meaning that the odds of a Yes vote do not change at all in response to changes to income. We should also check out the confidence intervals using exp(confint(chOut)):

```
               2.5 %       97.5 %
(Intercept)    0.3544246   0.6180933
age            1.0128365   1.0261271
income         0.9999975   1.0000020
```

These results jibe with the hypothesis tests: the confidence interval for income straddles 1:1, confirming the nonsignificant result for that coefficient. The 95% confidence interval for the intercept ranges from 0.35:1 upward to 0.61:1, but we won't bother to interpret it for the reasons noted above. The confidence interval for age runs from a low of 1.0128:1 up to 1.0261:1. You can scale this in your head if it helps: Let's think about the one-decade difference, say for example between a 30-year-old and a 40-year-old. At the low end of the confidence interval, an increase in age of 1 year corresponds to 1.0128:1 in favor of a Yes vote. That's 1.28% increase in the likelihood of Yes for each increase in age of 1 year. At the high end of the confidence interval, an increase of 1 year corresponds to 1.0261:1 in favor of a Yes vote. That's a 2.61% increase in the likelihood of Yes. Multiply each percentage by 10 to consider a whole decade of change in age. So, for a 10-year change in age, the confidence interval runs from 12.8% up to 26.1% increase in the likelihood of voting Yes.

If you find that logic a little squirrely, try rerunning the analysis after scaling the age variable to represent decades instead of years. You can use this command to accomplish that simple transformation: ChileYN$age <- ChileYN$age/10. In fact, anytime you are having difficulty making sense of an odds ratio, particularly a small fractional one, you can adjust the scaling of the predictors with a linear transformation to make the results more interpretable. You will find that the coefficients change, in sensible ways, but the results of the significance tests do not.

As before we should also calculate and report the chi-square test using the command anova(chOut, test="Chisq") to produce the following (slightly abbreviated) output:

Model: binomial, link: logit
Response: vote
Terms added sequentially (first to last)

	Df	Deviance Resid.	Df Resid.	Dev	Pr(>Chi)
NULL			1702	2360.3	
age	1	34.203	1701	2326.1	4.964e-09 ***
income	1	0.062	1700	2326.0	0.8032

We have separate tests that compare three "nested" models. The first chi-square test compares the null model to a model that just includes the age predictor. The second chi-square test compares the model with just age to a model that has both age and income as predictors. Only the first chi-square test is statistically significant (because $p = 4.964e-09$ is below the threshold of $p < .001$). This result makes sense in light of the significance tests on the coefficients and confirms the utility of a model that contains only age.

You will notice that in each successive line of the output, we lose a degree of freedom each time we enter a new predictor. Also note that the column titled, "Deviance Resid.", is the chi-square value for the effect of the predictor, while the column entitled, "Dev" is the chi-square that represents what is unaccounted for in the dependent variable after the entry of each predictor in the model. You can work your way forward or backward in the table: For example, 2360.3 (from the top line) minus 34.203 (from the second line) leaves 2326.1 (also on the second line). The 34.203 is a chi-square value which is tested for significance on one degree of freedom.

One last topic to consider is effect size. If you remember from ANOVA and regression, it is possible to get a standardized effect size value, either R-squared or eta-squared, that summarizes the overall goodness of the model being tested. The mathematics of R-squared are based upon the use of the least-squares criterion, and as we are not using that method with logistic regression, there's no way to directly calculate the plain, old R-squared that we are used to. Fortunately, statisticians through the ages have pondered this problem and come up with a variety of "pseudo-R-squared" measures that give us similar information as R-squared provides. Our colleagues at Baylor University have organized these into a package called "BaylorEdPsych," that we can install and run like this:

```
install.packages("BaylorEdPsych")
library(BaylorEdPsych)
PseudoR2(chOut)
```

In the output below, I show only a subset of the estimates produced by the PseudoR2() command and I have taken the liberty of transforming the scientific notation for clarity:

McFadden	Adj.McFadden	Cox.Snell	Nagelkerke
0.015	0.011	0.012	0.027

There is a substantial amount of debate in the statistical community over which pseudo-R-squared measure is the best one (Smith & McKenna, 2012). Each of the four shown above has slightly different properties and will produce slightly different values depending upon the nature of the data being tested. If you are reporting the results of your own research you should report more than one value. Many journal articles have included the McFadden pseudo-R-squared, the Cox-Snell pseudo-R-squared, or the Nagelkerke pseudo-R-squared. The Nagelkerke comes out consistently larger than the others and Smith and McKenna (2012) suggest that it is the closest analog to the plain, old R-squared that is used in least-squares models. For any of the given measures, you can *loosely* interpret it as the proportion of variance in the outcome variable (vote) accounted for by the predictor variables (income and age). In that light, these are all very small effect size values. Given our finding that only age was significant, these results suggest that age has only a rather small role in accounting for the vote variable.

You might ponder why age was statistically significant if its effect on vote is so small. This conundrum returns us to the idea of statistical power. Because we had a relatively large sample size of $n = 1,703$, we were able to detect a much smaller effect size with these data. As a reminder, Cohen (1992) published a great article entitled "A Power Primer" that delves into the meaning of statistical power and provides practical strategies for improving statistical power.

To close this section, it is common practice when testing categorical models to create a small table that compares the observed outcome to predicted results. Such a table is often referred to as a **confusion matrix,** presumably because it can show how confused the prediction model is. We can access the predicted values from the model output object with the command predict(chOut, type="response"). The type="response" argument to this command makes sure that the predicted results are calibrated in terms of the outcome variable. Remember that we are modeling the probabilities of a Yes or No response on the logit curve, so all of the predicted values will range between 0 (corresponding to No) and 1 (corresponding to Yes). Every response at or above 0.5 signals that the model is predicting a Yes vote, while everything below 0.5 signals a No vote. You could examine the distribution of these predicted probabilities by creating a histogram with hist(predict(chOut,type="response")). We will need to dichotomize those probabilities in order to create our confusion matrix. As we did earlier in the chapter, we can use the round() function to round every value lower than 0.5 down to 0 and all the others up to 1. Here is a command that accomplishes what we need:

table(round(predict(chOut, type="response")), ChileYN$vote)

In this table() command, we ask for the rows to be the predicted outcome values from the model, dichotomized to either be 0 (signifying a No vote) or 1 (signifying a Yes vote). In the columns we put the original vote values from our transformed data set ChileYN. The two-by-two contingency table that results from the table() command is here:

	N	Y
0	565	449
1	302	387

So the correct predictions (some analysts call them classifications) are on the main diagonal. To be specific, there were 565 cases where the observed vote in the data set was No and the dichotomized predicted value was 0. Those are the correct classifications of No. Likewise, there were 387 cases where the observed vote in the data set was Yes and the dichotomized predicted value was 1. The off-diagonal items, 449 and 302, are all of the erroneous predictions. You can easily calculate an overall accuracy of (565+387)/1703 = 0.56. The 1703 in the denominator is the total number of observations, which you can get either by looking back at the original data set or adding up the four cells. With these data you could just do a random bunch of guesses and over the long run get about 50% right, so an overall accuracy of 56% is not very good. This confirms what the other evidence (such as the pseudo-R-squared) showed us: although age was a significant predictor of Yes/No vote, a predictive model including income and age was not very good. By the way, you should look in the table and be able to identify the number of false positives (where the data said No, but the model said 1) and the number of false negatives (where the data said Yes, but the model said 0). Which type of error is more common with this model, false negatives or false positives?

BAYESIAN ESTIMATION OF LOGISTIC REGRESSION

Just as we could use Bayesian methods for ANOVA and linear regression, we can also use them for logistic regression. There are no major conceptual differences in the application of Bayesian methods to logistic regression versus the earlier techniques we have examined. The goal of the Bayesian analysis is to use a set of weakly informative priors concerning the coefficients to be estimated, and then generate a set of posterior distributions using the Markov chain Monte Carlo technique. The results include posterior distributions for each coefficient where the mean value becomes our point estimate of the coefficient, and the distribution around the mean shows us the highest density interval in which the population value of the coefficient is likely to lie. The friendly folks who wrote the BayesFactor package have not created a logistic regression procedure, but fortunately, Martin, Quinn, and Park (2011) created MCMCpack, an R package that allows us to accomplish similar goals. In fact, MCMCpack draws upon some of the same underlying packages as BayesFactor (such as the coda package), so you will notice some similarities in the output from the following commands:

```
install.packages("MCMCpack")     # Download MCMCpack package
library(MCMCpack)                # Load the package
ChileYN$vote <- as.numeric(ChileYN$vote) - 1   # Adjust the outcome variable
bayesLogitOut <- MCMClogit(formula = vote ~ age + income, data = ChileYN)
summary(bayesLogitOut)                   # Summarize the results
```

In the code above we downloaded the MCMCpack package from the Internet, then used the library() command to load it into memory. The third line of code makes a minor transformation of the outcome variable. Whereas glm() happily accepted a factor with two levels as the dependent variable, MCMClogit() expects the dependent variable to have numeric values of either 0 or 1. I performed a little code hack by using as.numeric() to convert the factor coding (which coded No = 1 and Yes = 2) into a number and then subtracting 1 to adjust the values. The fourth line of code runs MCMClogit() with the same model specification as we used for glm(). Finally, I ran summary() on the result, which produced the following output:

1. Empirical mean and standard deviation for each variable, plus standard error of the mean:

	Mean	SD	Naive SE	Time-series SE
(Intercept)	-7.602e-01	1.452e-01	1.452e-03	4.849e-03
age	1.935e-02	3.317e-03	3.317e-05	1.089e-04
income	-3.333e-07	1.151e-06	1.151e-08	3.723e-08

2. Quantiles for each variable:

	2.5%	25%	50%	75%	97.5%
(Intercept)	-1.044e+00	-8.588e-01	-7.609e-01	-6.639e-01	-4.710e-01
age	1.278e-02	1.711e-02	1.930e-02	2.157e-02	2.584e-02
income	-2.549e-06	-1.113e-06	-3.662e-07	4.437e-07	1.926e-06

As a reminder, this output focuses on describing the posterior distributions of parameters representing both the intercept and the coefficients on age and income, calibrated as log-odds. At the top of the output shown above, you can read the point estimates for these distributions under the "Mean" column. The point estimates for the intercept and the coefficients are quite similar to the output from the traditional logistic regression, although the mean value for the income coefficient is somewhat smaller. The next column to the right, labeled "SD," corresponds to the standard error in the output from the traditional logistic regression (because in this analysis it is a standard deviation of a sampling distribution). The second part of the output displays quantiles for each coefficient, including the 2.5% and 97.5% quantiles. The region in between the 2.5% and the 97.5% quantiles for each coefficient is the highest density interval (HDI) for the given coefficient.

We can get a more detailed view of these HDIs by using the plot function. The results of the following command appear in Figure 10.6:

plot(bayesLogitOut)

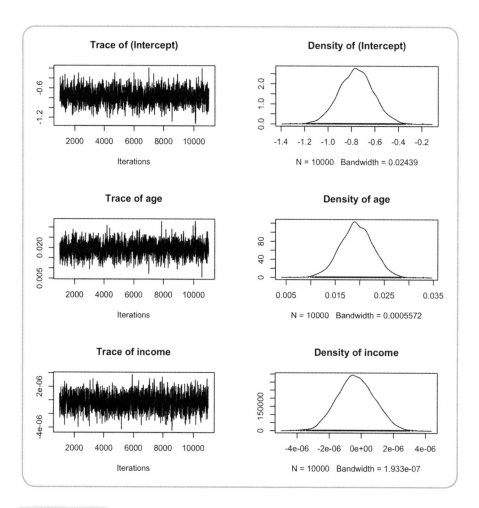

FIGURE 10.6. Highest density intervals of MCMC coefficient estimates from logistic regression (right-hand column).

The left-hand column of Figure 10.6 gives a "trace" of the progress of the MCMClogit() command as it conducted the Markov chain Monte Carlo analysis. For each of the 10,000 iterations of the algorithm, the height of the corresponding black lines shows the value of each coefficient for that iteration. The only problem that might occasionally appear in these traces is if the fluctuating pattern had a noticeable rise or drop over time, or if one end of the pattern was much more variable than the other end. If that occurs, you can sometimes stabilize the result by running more than 10,000 iterations. In this case, neither of those problems arose, so we can look over at the density plots in the right-hand column. These give a graphical representation of the likely position of each coefficient. The true population value of each coefficient is likely to be somewhere near the middle of each distribution and much less likely to be somewhere out in the tails. Note that the density plot for income clearly overlaps with 0, clarifying the evidence from the significance test that glm() provided for this coefficient. This point again illustrates the value of being able to graphically review the posterior distribution and the HDI, as it gives us extensive information about the "alternative hypothesis" for each of the coefficients being estimated.

It's a bit annoying, however, to have to review all of these distributions in terms of log-odds, so I wrote a few lines of code that will allow us to examine the posterior distribution of age in terms of regular odds. You could apply this same code to both the intercept and the income variable with the appropriate substitution in the first line of code:

```
ageLogOdds <- as.matrix(bayesLogitOut[,"age"])  # Create a matrix for apply()
ageOdds <- apply(ageLogOdds,1,exp)         # apply() runs exp() for each one
hist(ageOdds)                              # Show a histogram
abline(v=quantile(ageOdds,c(0.025)),col="black")   # Left edge of 95% HDI
abline(v=quantile(ageOdds,c(0.975)),col="black")   # Right edge of 95% HDI
```

In the first line, I coerce the age coefficient (in its log-odds form) into a matrix to get it ready to submit to apply(). Then I run apply() on that new data object with a function call to exp(). Using apply() in this way runs exp() repeatedly on each row in the ageLogOdds object, thereby converting each of the 10,000 values in the posterior distribution from log-odds to regular odds. I store the result in ageOdds, and then create a histogram with the edges of the 95% HDI marked off with vertical lines. The result appears in Figure 10.7.

Figure 10.7 shows a symmetric distribution centered on about 1.02, consistent with the results that we obtained from the glm() analysis and suggesting an increase of about 2% in the likelihood of a Yes vote for an increase in age of 1 year. While the most likely values of the coefficient in the population are in the area near that center point, the 95% HDI spans a region starting as low as 1.013 and ranging as high as 1.026. These boundaries for the HDI are similar, but not identical, to those of the confidence interval we obtained from the glm() analysis. As always, the histogram depicting the 95% HDI gives us a direct view of the most likely range of coefficient values in the population.

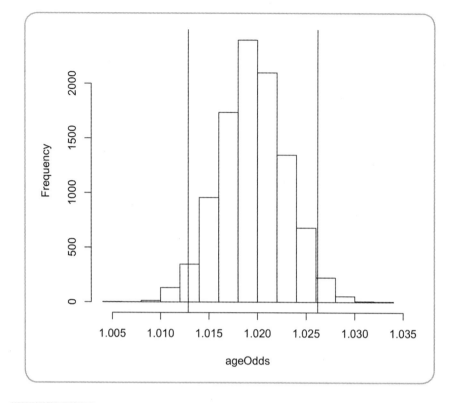

FIGURE 10.7. Histogram of the posterior distribution of odds for the age predictor variable.

CONCLUSION

In this chapter we extended our understanding of predictive models by examining models with binomial or dichotomous outcome variables. The logic of these models is similar to linear regression: we have a set of predictors that can include metric variables and categorical variables (although we only used metric predictors in the examples in this chapter) and we are trying to do the best job we can of predicting an outcome variable. The big difference is that in linear regression the outcome variable also has to be metric (and should be normally distributed) whereas in logistic regression the outcome variable can be binomial (e.g., yes/no, true/false). Some people refer to the prediction of a binomial variable as a "classification" analysis and in this sense logistic regression might be considered a kind of simplistic data-mining technique.

Modeling a true dichotomous output variable with statistics is difficult, so as a kind of shortcut, logistic regression tries to predict the Y variable as if its values fell on an inverse logit curve. An inverse logit curve is S-shaped and smooth—right in the middle of the curve it makes a rapid transition in slope that signals the change from one outcome state to the other. We use what is called the generalized linear model with a "link function" that involves the inverse logit curve. The outputs of this model include estimates of each coefficient on our predictor variables. These are reported as the "log-odds" such that a one-unit change in the predictor affects the log-odds of a state change of the outcome variable according to the coefficient. In almost all cases we want to convert the log-odds into regular odds so that we can interpret the results in a more sensible way.

Using the traditional methods of significance testing we can test the null hypothesis that any given coefficient is actually 0 in the population (in its log-odds representation), with the hope of rejecting that null hypothesis. We can also compare the chi-square value of a null model to the chi-square value of a model that contains one or more predictors to see if the reduction in error was significant. The null hypothesis here is that the addition of the predictor(s) did not improve the model. If the change in chi-square is significant, we reject the null hypothesis.

In contrast, the Bayesian method of conducting logistic regression gives us direct insight into the likely population value of each coefficient. By using the Markov chain Monte Carlo (MCMC) method, we can generate lots of alternative solutions that develop a posterior distribution for each coefficient. From the output we can obtain the HDI for each coefficient. We can feel certain that the population value of the coefficient lies somewhere close to the mean value of the coefficient generated by MCMC. When an HDI does not overlap 0 in either tail, we have provided evidence for an alternative hypothesis of a nonzero coefficient for a given predictor.

Throughout the book we have been concerned with the practical meaning of each statistical model that we consider and to this end we have reviewed various measures of effect size such as R-squared and eta-squared. In the case of logistic regression, we cannot calculate the same R-squared value that we got used to with ANOVA and regression, but we did find a variety of pseudo-R-squared values that give us similar information about the quality of our logistic regression models. With logistic regression there is also a concept of "goodness of fit" that is captured by the chi-square value or the AIC of a given model. Chi-square can be used to compare nested models (where one model is a subset of another model) and AIC can be used to compare nonnested models. Finally, the confusion matrix gave us an overall view of our prediction quality and allowed us to review the amount of false positives and false negatives.

EXERCISES

1. The built-in data sets of R include one called "mtcars," which stands for Motor Trend cars. Motor Trend was the name of an automotive magazine and this data set contains information on cars from the 1970s. Use "?mtcars" to display help about the data set. The data set includes a dichotomous variable called vs, which is coded as 0 for an engine with cylinders in a v-shape and 1 for so called "straight" engines. Use logistic regression to predict vs, using two metric variables in the data set, gear (number of forward gears) and hp (horsepower). Interpret the resulting null hypothesis significance tests.

2. Using the output from Exercise 1, convert the log-odds coefficients into regular odds and interpret them. Only interpret coefficients on predictors that were significant. Generate and interpret confidence intervals around the coefficients of the significant predictors. Would it make sense to rescale any of the predictor variables to make the results more interpretable? Why or why not?

3. Conduct the same analysis as in Exercise 1 using the MCMClogit() function provided in the MCMCpack package. Don't forget to use install.packages() and library() if this is the first time you have used MCMCpack. Report the point estimates (means) and 95% HDIs for the intercept and each predictor. Compare the results from the Bayesian analysis with the results from the traditional null hypothesis significance tests.

4. Plot the output from the MCMClogit() analysis in Exercise 3. Interpret the results of plots in your own words.

5. As noted in the chapter, the BaylorEdPsych add-in package contains a procedure for generating pseudo-R-squared values from the output of the glm() procedure. Use the results of Exercise 1 to generate, report, and interpret a Nagelkerke pseudo-R-squared value.

6. Continue the analysis of the Chile data set described in this chapter. The data set is in the "car" package, so you will have to install.packages() and library() that package first, and then use the data(Chile) command to get access to the data set. Pay close attention to the transformations needed to isolate cases with the Yes and No votes as shown in this chapter. Add a new predictor, statusquo, into the model and remove the income variable. Your new model specification should be vote ~ age + statusquo. The statusquo variable is a rating that each respondent gave indicating whether they preferred change or maintaining the status quo. Conduct general linear model and Bayesian analysis on this model and report and interpret all relevant results. Compare the AIC from this model to the AIC from the model that was developed in the chapter (using income and age as predictors).

7. Bonus R code question: Develop your own custom function that will take the posterior distribution of a coefficient from the output object from an MCMClogit() analysis and automatically create a histogram of the posterior distributions of the coefficient in terms of regular odds (instead of log-odds). Make sure to mark vertical lines on the histogram indicating the boundaries of the 95% HDI.

## CHAPTER 11

Analyzing Change over Time

All of the data sets we have considered so far have an interesting property that you might not have noticed: in these data, time was not a consideration in the measurement of the variables or the analysis of the data. The data we have considered so far were **cross-sectional,** meaning that all of the data were collected at (roughly) the same time. Scientists and statisticians use the term "cross-sectional" because each case or observation is like a small "slice" of the observed phenomenon. Think, for example, of the famous "iris" data set, where we ran a correlation analysis of sepal width and petal width. When Edgar Anderson (1935) was out in the field measuring iris plants, he collected all of the measurements from each plant at the same time.

In certain cases, time might have been implicit in the measurement process, but not in a way that allowed us to assess changes over time. For example, to measure fuel economy in the mtcars data set, testers had to drive each car over a period of time. Yet none of the information about the passage of time was included in the data set. Most importantly, we only had one measurement of fuel economy for each car.

Similarly, one of the earlier homework exercises asked you to analyze the "rock" data set that contained measures of permeability. One of the ways that you can measure permeability is to measure how long it takes a fixed amount of liquid to pass through a substance. But even though it sounds like the researchers measured time in that study, time was only a substitute or "proxy" for another kind of measurement. The researchers only measured the permeability of each rock sample once, so the passage of time and its possible influence on rocks was not considered. For both the mtcars and the rock data sets, because each case was only measured one time, we could not learn anything about what kind of changes or influences might have occurred over time. To understand changes across time, you must measure the same "thing" at least twice.

Many phenomena in business, education, government, social sciences, engineering, and natural sciences have an immensely important connection to time, so understanding what makes these data special and being able to analyze them properly is an essential skill for data analysts. In this chapter we will consider just a few of the different ways that one might analyze change over time, as well as some of the complications that arise when you do measure something more than once. In particular, we will consider two configurations of data that researchers use: **repeated measures** and **time series.** In a repeated-measures study, several subjects are measured at two or more points in time. In many cases the study includes some kind of intervention or activity in between different time points. A classic example would be a study of learning, where individuals take a test prior to an educational experience, then they have that experience, and then they are tested again. Repeated-measures studies usually capture only a few different points in time, as a rule of thumb no more than about a dozen.

In contrast, a time-series study usually measures a single phenomenon or a small number of closely related phenomena over many time intervals. One of the most commonly used examples of a time series is the price fluctuation of a stock market. Depending upon the analytical technique that is used, a time-series study usually requires at least a few dozen data points. Generally, the most useful time-series data sets tend to contain many hundreds or even thousands of data points. It is also a typical, though not universal, requirement that time-series data are collected at equal intervals. So if you collected one data point at 2:00 P.M. and another at 2:01 P.M., then the next data collection would most certainly need to be at 2:02 P.M.

One of the most essential concepts to grasp about data collected over time is that observations are not **independent.** In nontechnical terms, when two observations are independent, it simply means that there was no connection between the source of variance that created one observation and the source of variance that created the other observation. In contrast, when two observations are dependent, this means that they are somehow connected such that we can expect them to be related and possibly correlated.

You can do an easy mental test of dependency by asking yourself what you may know about any observation B based on knowledge of another observation A. Let's say that observation A is the height of a randomly selected boy and observation B is the height of another randomly selected boy. These observations are independent because the value of A tells you nothing about the value of B. Now let's change it up and say that observation A indicates that a particular boy was 6 feet tall at age 17. Then a second observation, B, is taken when that *same* boy is age 18. You're certain that B will show that the boy is at least 6 feet tall and possibly a little more, so observations A and B are not independent. If we had many pairs of such observations—one element of each pair measured at age 17 and one at age 18—it is quite likely that the earlier and later observations would be highly correlated.

With the cross-sectional data sets considered earlier in this book, there was an assumption that each observation was independent of all the others. With repeated-measures and time-series data, we have pairs or larger sets of observations that are connected. When we choose the appropriate analytical techniques for these data we gain some benefits from the connectedness of the data and we avoid some pitfalls—especially violations of the assumption of independence that go with most of the analysis methods we have considered so far.

REPEATED-MEASURES ANALYSIS

The simplest form of a repeated-measures study occurs when we collect information about each case or subject at precisely two points in time. As a very simple example, we could measure the resting heart rate of five individuals who come to the gym: call these A1, B1, C1, D1, and E1. Then we could have each person walk on a treadmill for 5 minutes and then measure their heart rates again: call these A2, B2, C2, D2, and E2. We now have a total of 10 data points, although we only have five subjects/cases. Most importantly, each pair of observations, for example, A1 and A2, are *not* independent because they were both measured from the same source of variance, subject A.

This kind of simple situation, with just two data points per person, brings us back to our old friend the Student's *t*-test. William Sealy Gosset (AKA Student) recognized an important difference between two independent sets of cases versus one set that was measured twice. In the latter situation, one could capture the amount of change between the first measurement and the second measurement simply by subtracting the first set of observations from the second set.

There's an important advantage in being able to do this subtraction. Let's consider a tiny data set with just two subjects. Each subject is practicing the art of juggling, but one subject is much more experienced than the other. At the beginning of the practice period, Subject X can juggle six flaming torches at once, while Subject Y can only juggle two. At the end of the practice period, Subject X can now juggle seven flaming torches while Subject Y can juggle three. The most natural way to look at this situation is that, on average, each subject improved by one torch after practicing, a very solid and promising result. We got that result by subtracting time 1 results from time 2 results: X2 − X1 = 1 and Y2 − Y1 = 1. We can do this subtracting because we know that the two groups of observations are dependent on each other.

Imagine now if we had treated the time 1 observations as independent from the time 2 observations—pretending that they were two completely different groups of jugglers. The time 1 "group" would have a mean of $(6+2)/2 = 4$ and the time 2 "group" would have a mean of $(7+3)/2 = 5$, again a mean difference between groups of one torch. But here's the big deal: the sample standard deviation within each of these groups is 2.83: a huge amount of within-group variance compared with the amount of change. So much within-group variance, in fact, that we would not be able statistically to detect the change in the number of juggled torches before and after practice unless we had a much larger sample size. The within-group variance we observed here is a function of individual differences—one very experienced juggler and one quite inexperienced juggler. With a repeated-measures design, we can eliminate the influence of those individual differences by only examining the amount of change for each individual rather than the overall differences between groups.

Let's take a look at some real data to illustrate these ideas. The built-in data set "ChickWeight" contains measurements of the weights of chicks for different numbers of days after hatching. In Figure 11.1, I used the command boxplot(weight ~ Time, data=ChickWeight) to generate a box plot of the weights of chicks at different points in time.

Figure 11.1 clearly shows a pattern of growth over time as indicated by the medians (thick black line in the center of each box) of chick weights at times ranging from 0 days after hatching to 21 days after hatching. Note, however, that there is also a good deal of variability, particularly in day 6 and beyond. To illustrate the dependent measures *t*-test, we will compare weights on two neighboring days, day 16 and day 18. In this data set, chicks were measured on even numbered days since hatching. We will run an independent samples *t*-test on these data, even though that is inappropriate, and compare those results to a dependent samples *t*-test. Here is some code to set up the data that we will use:

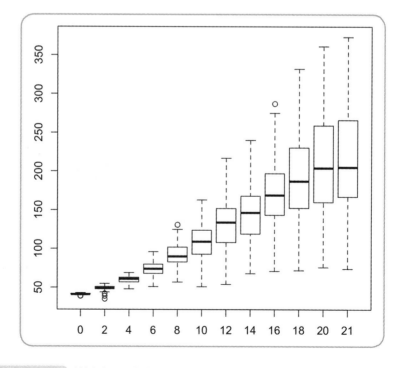

FIGURE 11.1. Weights of chicks in grams versus days after hatching.

```
ch16index <- ChickWeight$Time == 16      # Chicks measured at time 16
ch18index <- ChickWeight$Time == 18      # Chicks measured at time 18
bothChicks <- ChickWeight[ch16index | ch18index,]    # Both sets together
# Grab weights for t=16
time16weight <- bothChicks[bothChicks$Time == 16,"weight"]
# Grab weights for t=18
time18weight <- bothChicks[bothChicks$Time == 18,"weight"]
cor(time16weight,time18weight)                          # Are they correlated?
```

In the first line of code, we generate a list of all of the cases that have a measurement at time 16. In the second line of code, we generate a list of all of the cases that have a measurement at time 18. In the third line we use these lists to create a combined data set of the cases with measurements at time 16 and 18. The upright bar character in the row index of the third line is a "logical or" operator, meaning that we include cases that satisfy either condition. Because the *t*-test expects two vectors of measurements, the fourth and fifth lines subset the weight data from among the other columns in the data set. As a diagnostic, the final line examines the correlation between the two vectors of measurements. Do you expect that correlation to be high or low?

In fact, the correlation is about $r = 0.97$, a high correlation indeed. This high correlation reflects the extent and stability of individual differences among chicks: a chick that is small at age 16 days is still pretty small (though possibly slightly larger) at age 18 days. Likewise, a chick that is large at age 16 days is probably even larger at age 18 days. This dependency over time is the essence of why we want to use a repeated-measures analysis, as you will see below.

Let's first analyze these data as if they were two independent groups. Don't forget that this is, in fact, an incorrect approach to analyzing these data that I am undertaking in order to show the difference between independent groups and dependent groups analysis:

```
mean(time16weight)
mean(time18weight)
# Independent groups t-test
t.test(time18weight,time16weight,paired = FALSE)
BESTmcmc(time18weight,time16weight)       # Run the Bayesian equivalent
```

The first two commands reveal that the mean weight at 16 days is 168.1 grams, while the mean weight at 18 days is 190.2 grams. This is a promising mean difference, but we need to use a *t*-test to see if the difference in means exceeds what can be expected as a result of sampling error. The t.test, with the argument paired=FALSE, treats the two groups as independent, calculates a separate variance for each group, and then pools the variance to calculate the *t*-test. The result shows $t = 2.05$ on 88.49 degrees of freedom. This result is significant at $p < .05$, but the confidence interval, which ranges from 0.62 to 43.6, indicates that on the low end the mean difference is barely larger than 0. In the same vein, the Bayesian *t*-test shows that the 95% HDIs for the two means actually overlap, suggesting that there is not a credible difference between these two means when each group is considered as an independent entity.

Now let's perform the appropriate analysis, by correctly treating these two groups as dependent. The following command performs the traditional dependent groups *t*-tests.

```
t.test(time18weight,time16weight,paired = TRUE)    # Dependent groups t-test
```

In this command, we have changed the argument to paired=TRUE in order to get the t.test() function to treat these two vectors of data as a matched pair. The output shows several differences from the *t*-test output we have reviewed before:

```
Paired t-test
data: time18weight and time16weight
t = 10.136, df = 46, p-value = 2.646e-13
alternative hypothesis: true difference in means is not equal to 0
95 percent confidence interval:
17.71618 26.49658
sample estimates:
mean of the differences
22.10638
```

Note at the top of the output the designation "Paired t-test," confirming that the data have been treated as matched pairs, and the confirmation on the next line of the two variables that formed the matched pair. On the following line, the *t*-value of 10.1 is reported on 46 degrees of freedom, $p < .001$. This is a much larger value of *t* than what we observed in the independent samples *t*-test, but the real proof is in the confidence interval. This confidence interval shows a lower boundary mean difference of 17.7 and an upper boundary mean difference of 26.5. This is a much narrower confidence band than what we observed before and more importantly, the lower bound is nowhere near 0. A narrower band means greater certainty about the result and having the lower bound so far from 0 gives us clarity about the likely difference between day 16 measurements and day 18 measurements.

You should ask yourself why these results are so much stronger than when we (incorrectly) applied the independent samples *t*-test to the data. The explanation lies in that correlation of $r = 0.97$ between the day 16 and day 18 measurements. That correlation is high because of the individual differences among chicks and the stability of those individual differences over time. When we use a dependent samples test, we eliminate those individual differences, leaving only the amount of change for each case. We may have small, medium, and large chicks, but once we remove those individual differences what we are left with is the growth that occurs for each chick between day 16 and day 18. This idea is easy to demonstrate through the use of difference scores. In these next lines of code, I calculate a difference score for each chick, and then subject this vector of difference scores to both a traditional and a Bayesian analysis. In each case we are trying to ascertain whether, on average, the difference scores are meaningfully larger than 0:

```
weightDiffs <- time18weight - time16weight     # Make difference scores
t.test(weightDiffs)     # Run a one sample t-test on difference scores
# Run the Bayesian one-sample test on difference scores
BESTmcmc(weightDiffs)
```

When you run the code for the t.test() on the difference scores—technically called a "one sample" *t*-test—you will find that the output is functionally identical to what we just produced for the paired sample *t*-test:

```
One Sample t-test
data: weightDiffs
t = 10.136, df = 46, p-value = 2.646e-13
alternative hypothesis: true mean is not equal to 0
95 percent confidence interval:
17.71618 26.49658
sample estimates:
mean of x
22.10638
```

Interestingly, the degrees of freedom is also identical to the earlier analysis at df = 46. This value for degrees of freedom is now quite easy to understand: there are 47 difference scores in the weightDiffs vector and we lose 1 degree of freedom when we calculate the mean of the difference scores, leaving 46 degrees of freedom. One other thing to notice about this output is the statement on the third line: "alternative hypothesis: true mean is not equal to 0." If you give this some thought, you will see how the paired *t*-test and the one sample *t*-test are really exactly the same thing: by examining the difference for each pair of scores, we can figure out whether there has been change over time across the weights of the 47 chicks that were measured on day 16 and day 18. By pairing up the two sets of measurements we are correctly treating them as one group of chicks at two different points in time.

The output of the BESTmcmc(weightDiffs) command confirms what we learned from the one sample *t*-test. The output provides an HDI for the mean difference in measurements between day 16 and day 18 and shows a lower bound of 17.5 and an upper bound of 26.5—nearly identical to the confidence interval produced by the one sample *t*-test. Don't forget that the confidence interval is an interval estimate that shows one possible band of values that may or may not contain the actual population value. On the other hand, the 95% HDI displays the results of a sampling process from the posterior distribution of the population parameter (using the Markov chain Monte Carlo technique)—in this case the mean difference between measurements taken at the two points in time. While it is comforting to know that the confidence interval and the

95% HDI closely match each other, it is important to keep in mind that they represent different conceptual approaches to inference from the sample data.

Your understanding of data is sufficiently advanced at this point that you have probably realized an important shortcoming in our use of the *t*-test. Figure 11.1 shows that the ChickWeight data contains measurements at 12 different points in time. Yet the dependent/paired samples *t*-test is limited to comparing one time point to just one other time point. While in theory we could use the dependent samples *t*-test to compare each pair of neighboring measurements, this quickly becomes cumbersome. Of additional concern, when we are conducting many significance tests on the same data set, we must guard against finding false positives simply by chance. To transcend these limitations, we need to move from *t*-test to ANOVA, just as we did earlier in the book. Repeated-measures ANOVA allows us to compare cases across two or more points in time, rather than just one pair of points.

We can continue to use the ChickWeight data, but we must first condition it to establish what statisticians call a **balanced design.** Having a balanced design means that there is a measurement for each chick at each time point. If you examine the ChickWeight data closely, for example by using the table (ChickWeight$Chick,ChickWeight$Time) command, you will find that not all chicks have measurements at all times. In fact, chick number 18 only has two out of the 12 measurements. While there are more sophisticated analytical techniques that can handle this kind of imbalance, our old friend aov() works best when presented with a balanced design.

We can use a little bit of R programming to eliminate the cases that are missing one or more measurements:

```
chwBal <- ChickWeight                          # Copy the dataset
chwBal$TimeFact <- as.factor(chwBal$Time)       # Convert Time to a factor
# Make a list of rows
list <- rowSums(table(chwBal$Chick,chwBal$TimeFact))==12
list <- list[list==TRUE]                # Keep only those with 12 observations
list <- as.numeric(names(list))         # Extract the row indices
chwBal <- chwBal[chwBal$Chick %in% list,]      # Match against the data
```

This code looks complex but only introduces one or two things we haven't tried before. In the first line we simply copy the data set into a new object to leave the original pristine. The Time variable is a number in the original data so we convert to a factor to make the ANOVA work properly. The third line uses table() to make a summary of all of the observations and then boils this into a list or TRUEs and FALSEs: TRUE if all 12 observations are there and FALSE if not. The fourth line discards the FALSEs because these are the chicks that are missing observations at one or more points in time. The fifth line extracts the chick ID numbers from the list and the final line uses that list to keep the chicks with complete data. The last line uses the %in% operator, which is really cool. The expression "chwBal$Chick %in% list" means that we only keep the rows where the value of Chick for that row is somewhere in the list.

You can check the results yourself by generating another table with the command table(chwBal$Chick,chwBal$TimeFact). This should produce a table of Chicks by Time that has no empty cells (missing observations). The following code runs an ANOVA test on the resulting balanced data set that examines the effects of Time (as a factor) on weight:

```
summary(aov(weight ~ TimeFact + Error(Chick), data=chwBal))
```

Repeated measures ANOVA reflects much of the logic we discussed above. In the model formula "weight ~ Time" we are testing the hypothesis that weight does not vary over time. Each point in time can be considered a group, similar to the categorical factors that we previously examined with oneway ANOVA. Note, however, that we now have a new term in the ANOVA model—"Error(Chick)"—that specifies the individual differences among chicks as error variance that we want to leave out of out F-test. By specifying this error term, we are able to remove the confounding effects of individual differences from the overall error term, thus making it more likely that we will be able to detect change over time. The results of this call to aov() appear below:

Error: Chick

	Df	Sum Sq	Mean Sq	F value	Pr(>F)
Residuals	44	429899	9770		

Error: Within

	Df	Sum Sq	Mean Sq	F value	Pr(>F)
TimeFact	11	1982388	180217	231.6	<2e-16 ***
Residuals	484	376698	778		

This may look a little confusing but does have a sensible internal logic to it. If you add up all of the degrees of freedom, the total is 539. There are 540 total observations in this data set, and once we have calculated the grand mean, there are $df = 539$ remaining. Likewise, if you add up all of the sums of squares you will find that the total is 2,788,985, which is identical (within rounding error) to the total sum of squares in the whole data set (which you could calculate by multiplying var(chwBal$weight) by 539). What the ANOVA table shows, then, is that the variance and the degrees of freedom have been *partitioned* into various components, in pursuit of the goal of separating out the individual differences variance from the calculation of the F-test.

In the first section of the output, marked "Error: Chick," the "Residuals" refers to the variance attributable to individual differences among chicks. The $df = 44$ signifies that we have 45 chicks in the data set. The sum of squares of 429,899 represents variation in the weight variable that is directly attributable to individual differences among chicks. The main virtue of repeated measures ANOVA is that we can separate out this portion of the variation in the dependent variable and set it aside so that it does not appear in the denominator of our F-test.

Using ezANOVA

The aov() procedure offered in the base statistics installation of R does many things simply and well. In this chapter we used aov() to analyze a repeated-measures design and we worked very hard to make sure to provide a balanced data set to the procedure (i.e., all chicks have data at all times). In response, aov() provided output containing the basic ANOVA table. If we had failed to provide a balanced data design, however, aov() would have still provided output—but it would have been difficult to interpret (and technically incorrect, some statisticians argue). So with aov() we get bare-bones output and the possibility of obtaining incorrect results without warning.

In response to these shortcomings, statisticians and programmers have worked on a variety of more sophisticated alternatives to aov() that provide richer output and a broader range of diagnostics to help analysts avoid missteps. One such alternative is the ez package and its main function called ezANOVA(), written by Mike Lawrence (Lawrence, 2013). Here's the same repeated measure ANOVA that we just conducted with aov() but this time with the ezANOVA() command:

```
library("ez")
install.packages("ez")
ezANOVA(data=chwBal,dv=.(weight),within=.(TimeFact),wid=.(Chick),
    detailed=TRUE)
```

That code produces the following output:

```
$ANOVA
    Effect    DFn DFd    SSn      SSd        F        p p <.05        ges
1 (Intercept)  1   44  8431251 429898.6 862.9362  1.504306e-30 * 0.9126857
2 TimeFact    11  484  1982388 376697.6 231.5519  7.554752e-185 * 0.7107921
$'Mauchly's Test for Sphericity'
    Effect                W            p p <.05
2 TimeFact      1.496988e-17  2.370272e-280 *
$'Sphericity Corrections'
    Effect        GGe   p[GG] p[GG] <.05      HFe   p[HF] p[HF] <.05
2 TimeFact 0.1110457    7.816387e-23 * 0.1125621    4.12225e-23 *
```

In the first section, underneath the label \$ANOVA, the first line contains the calculation of individual differences. The second line provides the conventional F-test that matches what we found with aov(). The test, $F(11,484) = 231.6$, is statistically significant, because $p < .001$, so we can reject the null hypothesis of no change in weight over time. The final value on the second line, 0.71, is the "generalized" eta-squared, in this case identical to what we calculate on the next page. The next section, labeled "\$Mauchly's Test for Sphericity," tests for homogeneity of variance of the differences among pairs of time groups. Homogeneity of variance is

(continued)

an important assumption of repeated measures ANOVA and the fact that this test is significant means that we have violated the assumption. If you look back at Figure 11.1 you will see quite clearly that variance in weight is minimal when chicks are younger but increases gradually in each time group. When Mauchly's test of sphericity is significant, it raises the possibility that the *F*-test is incorrect and we may draw the wrong conclusion from the null hypothesis significance test on F.

Fortunately, the ezANOVA() command also provides some additional tests in the third section of the output that we can consult when we have violated the test of sphericity. The Greenhouse–Geisser (1959) correction, abbreviated above as GGe (with the associated probability $p[GG]$), applies a correction to the degrees of freedom to counteract the possible inflation of the *F*-ratio. The associated probability value, $p[GG]$, reflects the same *F*-ratio evaluated against a more stringent set of degrees of freedom. If the associated *p*-value remains significant, we do not need to revise our decision about rejecting the null hypothesis. The Huynh–Feldt (1976) correction, abbreviated above as HFe (with the associated probability $p[HF]$), also applies a correction to the degrees of freedom to counteract the possible inflation of the *F*-ratio. Of the two tests, Greenhouse–Geisser is considered more conservative. In this case, because both corrections show the same result, we can be confident that any inflation in the *F*-ratio, $F(11,484) = 231.6$, has not adversely affected our decision to reject the null hypothesis.

In the output shown above, we didn't experience any of the warning and error messages that ezANOVA() spits out if it believes you have not prepared the data properly. The ezANOVA() command does careful checking of the input data to help ensure that it does not produce confusing or incorrect results. In particular, ezANOVA() will indicate an error if you try to submit an unbalanced data set to repeated measures analysis.

There's lots more to learn about the ezANOVA() command, and the ez package also offers a variety of plotting and diagnostic functions that you may find very handy. As always, R offers help for the ez package and its commands once the package is installed and loaded. If you search online for examples and tutorials, make sure you use a search string like "ez package R," because there is another software program called ezANOVA that is not affiliated with the R package.

In the second section, marked "Error: Within," the effect of Time is expressed as an *F*-ratio, $F(11,484) = 231.6, p < .001$. The $df = 11$ for the numerator reflects the 12 points in time where we measured the weights for each of the 45 chicks. The $df = 484$ for the denominator is the remaining error variance that is not attributable to individual differences. This *F*-ratio tests the null hypothesis that *changes* in chick weights are consistently 0 across all time intervals. With a low *p*-value that is below the conventional threshold of significance we can reject this null hypothesis and thereby lend support to an alternative hypothesis that the weights of chicks do vary across different points in time. If you reexamine Figure 11.1, you will find that that there is a gradual increase in the median chick weight at each successive time interval. We can also calculate an eta-squared effect size for the influence of the Time variable using the sums of squares from the ANOVA table: $1982388/(1982388+376698+429899)$ $= 0.71$. We interpret this value as the proportion of variance in weight that is accounted for by Time, keeping in mind that we treated Time as a categorical factor.

We have just barely scratched the surface here with respect to both the power and the complexity of repeated-measures ANOVA. The important thing to keep in mind is that, by measuring the same "subjects" more than once, you can increase your statistical sensitivity for finding meaningful results. Both the dependent samples t-test and the repeated-measures ANOVA provide the opportunity to remove the pesky influence of individual differences before conducting the analysis of how time and other factors may influence a dependent variable. Although there is no strict limit on the number of points in time that one can examine using repeated-measures analysis, in most data collections researchers measure just a few points in time. A typical study might capture a baseline measurement, then apply an intervention, then measure change at a second point in time, and finally undertake a follow-up after a more extended period of time to see how effects persist. In our examination of the chick weights data we also saw how researchers repeatedly measured a variable for several "subjects" or cases over 12 points in time. Repeated-measures ANOVA was useful for analyzing this kind of data as it allowed us to partition and remove the individual differences among the chicks.

TIME-SERIES ANALYSIS

As an alternative approach to repeated-measures designs, one might choose to capture and analyze many measurements of one subject or phenomenon, an approach that researchers call **time-series analysis.** Whereas in the Chick-Weights data we measured the growth of about 45 different chicks at 12 points in time, in a time-series study we might measure one chick at 100 or more points in time. During that more lengthy period of time, we might be able to detect one or more events or anomalies that happen to the subject, such as a change in environment or the use of a drug or other intervention. Whereas repeated-measures analysis allowed us to consider the effects of individual differences, time-series analysis gives us the ability to examine and control for the effects of trends and cycles.

Trends and cycles in time-series data can be interesting in and of themselves but can also interfere with our ability to examine other relationships that may exist in the data. This conundrum is easy to illustrate with a few lines of R code and some line graphs. First, let's create a simple growing time series that we can pretend is a measurement that grows over time such as the weight of a chick or the change in price of some product or commodity:

```
set.seed(1234)                          # Control random numbers
tslen <- 180                            # About half a year of daily points
ex1 <- rnorm(n=tslen,mean=0,sd=10)      # Make a random variable
tex1 <- ex1 + seq(from=1, to=tslen, by=1)   # Add the fake upward trend
plot.ts(tex1)          # Plot the time series with a connected line
```

As usual, we begin with set.seed() so that your sequence of random numbers and mine are the same. Use a different seed value or take this line out if
you want to generate a truly random result. Next, we create $n = 180$ random
data points in roughly a normal distribution—if we think of each data point
as 1 day as we did with the ChickWeight data, this is about half a year of data
points. I have chosen a standard deviation of 10 just so that we get a nice jagged result. Finally, we can make this into a plausible but fake time series with a
growth trend just by adding an increasing integer to each subsequent value. In
the last line of code we order up a sequential plot of the data points. By using
a specialized plot command, plot.ts(), we get a line connecting the data points
rather than a series of individual dots. This is a standard convention in plotting
time-series data: if the data points represent a time sequence, either connect
the dots or show the result as a continuous line. The resulting graph appears in
Figure 11.2.

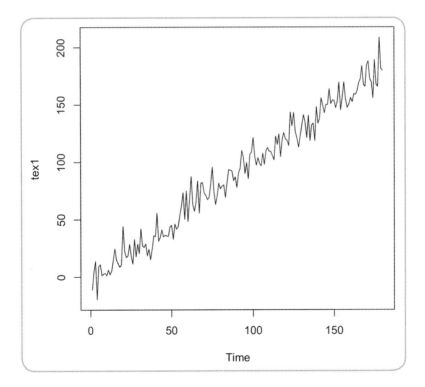

FIGURE 11.2. Time-series plot of $n = 100$ fake data points with upward
trend.

You can see from Figure 11.2 why we can think about this as the growth of a chick or an increase in prices. In general, over time, these kinds of measurements increase in magnitude: we can see that trend in the plot. On the other hand, from day to day there is some degree of variation up and down: that's why the plot looks jagged. Next, to illustrate an important analytical challenge, let's make a second variable using a new set of random numbers and then correlate the first variable with the second variable. Here's the code to create the second random variable:

```
ex2 <- rnorm(n=tslen,mean=0,sd=10)        # Make another random variable
tex2 <- ex2 + seq(from=1, to=tslen, by=1)      # Add the fake upward trend
cor(ex1, ex2)        # Correlation between the two random variables
cor(tex1, tex2)       # Correlation between the two time series
```

Again we start by creating a random variable with a normal distribution and a standard deviation of 10. We can pretend that this is another growth indicator, for example, spot prices on a commodities market. We use the same trick as before to add an upward trend. You can and should make your own time-series plot of the new variable "tex2" just to verify that it has similar characteristics to "tex1," but is nonetheless a distinctive data set. When we correlate the two original random normal variables using cor(ex1,ex2), we get a tiny correlation of $r = -.094$, just as we would expect. These two variables were each created as random normal distributions and should be effectively uncorrelated. But when we correlate the two fake time-series variables we get a monstrously huge correlation that is nearly $r = .96$. The time-series data are highly correlated *because of the upward trend that we inserted in both time series*.

You are probably thinking, "Duh," at this point, but consider if you had just received the two variables, tex1 and tex2, with no explanation other than the idea that they were meant to be two different indicators of some phenomenon. If you had simply done a scatterplot of the two variables together you would have seen what appears in Figure 11.3.

What would you have concluded? If I had seen this scatterplot with no other explanation, I would have said, "Wow, here we have two highly correlated variables!" If I then learned that these were two indices of economic activity, I would then have concluded that the two indices were measuring essentially the same phenomenon—for example, two different measures of the same stock market. Yet we know that the only reason that these two variables are so highly correlated is that they both contain the same growth trend. The beautiful linear shape of this scatterplot is an artifact that comes from the growth trend. Of course the economy does tend to grow over time, but that fact alone does not make the two variables measurements of the same thing. In fact, you could even think of the growth trend as something of a nuisance effect. We generally want to remove the trend if we plan to analyze these data using correlational techniques so that we do not make a mistake and treat the trend as evidence of a real connection between the two variables.

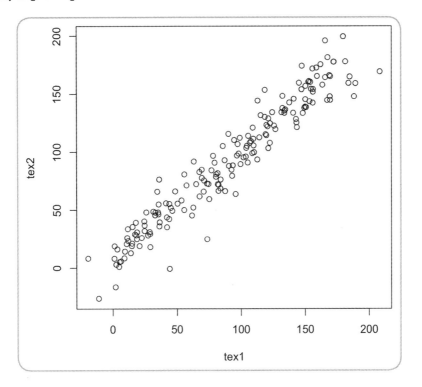

FIGURE 11.3. Misleading bivariate scatterplot of two uncorrelated time series that both contain a trend component.

In fact, that is one of the important applications of time-series analysis. We can decompose a time series into its component parts, so that we can examine each part separately and think about them separately from an analytical standpoint. We can pay attention to the trend and other aspects of the time series if we want to, or we can ignore them if we decide to look at some other aspect of the time series.

In fact, most statisticians think of a time series as being made up of four distinctive components. We have already examined **trend,** one of the most noticeable and obvious aspects of a time series: growth or decline across time. Another component of a time series is **seasonality:** regular fluctuations that occur over and over again across a period of time. The word "seasonality" is intentional and reflects the fact that some time series, when measured over the course of a year, have typical fluctuations up or down for spring, summer, fall, and winter. Closely related to seasonality is **cyclicality:** the idea that there may be repeating fluctuations that do not have a regular time period to them. Using another economic example, recessions come and go over periods of several years, but there is no precise regularity to them. The next recession may be 3 years away or 5 years away—we don't know for sure how long, but we can be certain that there will be another one at some point. Finally, each time series has an **irregular** component. Some statisticians refer to this component as **noise,** which is okay from a technical standpoint but confusing from a practical standpoint: the irregular component is often the main thing that we want to analyze. So calling the irregular component "noise" is somewhat confusing, as noise is something that we usually want to get rid of.

So, for many purposes we can think about time-series analysis as a process of trying to find out what is happening with trend, seasonality, cyclicality, and irregularity; to break apart those components into separate pieces; and then to analyze the pieces as we would any other quantitative data. To illustrate this, let's fiddle around with one more fake variable before we turn to some real data:

```
ex3 <- rnorm(n=tslen,mean=0,sd=10)
tex3 <- ex3 + seq(from=1, to=tslen, by=1)       # Add the fake upward trend
tex3 <- tex3 + sin(seq(from=0,to=36,length.out=tslen))*20
plot.ts(tex3)
```

Here we create another random normal variable and as before we add a trend to it using a simple seq() function to create a sequence of increasing integers. For a final flourish, however, we add a seasonal component to the data in the form of a sine wave. The sine function is the most basic of the trigonometric functions and here we have created roughly six cycles (one cycle is two times pi, or a little more than 6). The resulting time-series plot appears in Figure 11.4.

In addition to the growth trend that we saw before and the irregular ups and downs, Figure 11.4 shows six discernible cycles in the data. If these were real data, rather than fake data, our next task would be to decompose the time series into its component elements. Not unexpectedly, R has a function that can do this job for us.

```
decOut <- decompose(ts(tex3,frequency=30))
plot(decOut)
```

The decompose() function takes as input a time-series object. I converted tex3, which was a simple vector of values, into a time-series object using the ts() function. One of the parameters to the ts() function is "frequency." The frequency parameter sets the context in which the time-series analysis takes place—in other words the natural time scale on which seasonal variation may occur. In this case setting frequency = 30 signifies that we are treating each element in our time series (in this example, a day) as part of a larger pattern, namely 30 days per month (approximately speaking). Not all kinds of time-series analysis require this kind of contextual input, but in this case because we are trying to detect seasonality with the decompose() function, we must specify a larger aggregate of time over which repetitious patterns may emerge.

You may find this concept easier to understand by considering the results of the plot() command in the last line of code shown previously. The command, plot(decOut), produces the decomposition plot shown in Figure 11.5.

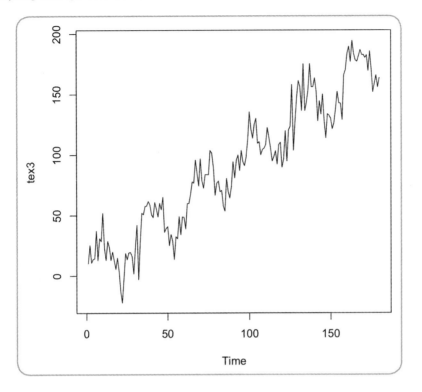

FIGURE 11.4. Time series that contains a trend component and a time-series component.

Note that the *X*-axis in Figure 11.5 is calibrated from 1 to 7, and this represents the range of months (i.e., collections of 30 days) in our data set of 180 days.

In Figure 11.5 there are four panes stacked on top of one another. In the top pane, there is a time-series line chart of the original series. This is the same as Figure 11.3, albeit with the *Y*-axis quite squashed. Next, the trend shows the linear climb that we introduced to the time series using the seq() command. The seasonal pane shows the periodic fluctuation that we put into the series using the sin() function to create an oscillating sine wave. It is important to look closely at this component of the time series to note that the re-creation of the sine wave is imperfect. The algorithms inside the decompose() function look for seasonal oscillations on the time scale provided by the frequency parameter that was used to create the time series (in this case frequency = 30). These algorithms cannot do a perfect job because of the irregular component of the time series: it is easy to mistake random fluctuations in the irregular component for seasonal fluctuations. This point underscores the importance—when creating a time-series data object—of correctly specifying the "natural time scale" of the time-series sequence with the frequency command. In some analytical situations the natural time scale of the data will be obvious—for

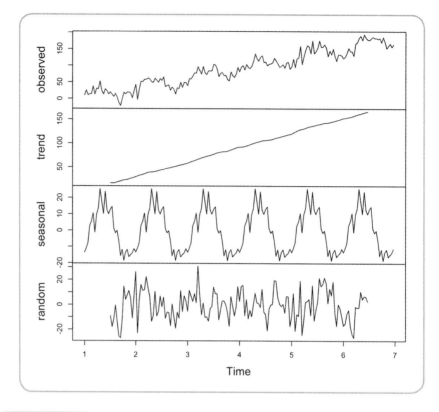

FIGURE 11.5. Additive decomposition of fake time series containing a seasonal component.

example, tides typically vary on a scale of hours whereas seasons change on a scale of months. In other analytical situations you may have no information on the time scale of variation, in which case you will have to experiment with different frequency options. In our fake example here, we used frequency = 30 to indicate that the times series contained 30 equally spaced samples within a month (and by extension, having $n = 180$ total observations means that we had exactly 6 months).

The bottommost pane in Figure 11.5, labeled "random," contains the irregular or noise component of the time series. This is all the variation that is left over after the trend and the seasonality are subtracted out of the time series. In the following code, I run some diagnostics that help us understand the contents of the resulting decomposition data object. We can access all of the data that were plotted in Figure 11.5 by looking inside the decOut object:

```
mean(decOut$trend, na.rm="TRUE")
mean(decOut$seasonal)
mean(decOut$random,na.rm="TRUE")
cor(ex3, decOut$random, use="complete.obs")
```

In the first line, we request the mean of the $trend element of decOut. The mean, equal to 90.2, perfectly reflects how we created the trend using a sequence of numbers from 1 to 180. In the second line we request the mean of $seasonal, which, as we saw from Figures 11.4 and 11.5, is a sine wave (with some random artifacts) that oscillates around 0. As a result, the mean of this component is 1.88e-16, a value very close to 0. Finally, the remaining component is what statisticians call the "irregular part" of the time series. The mean of this component, which is accessed with $random, is −0.57, just slightly below 0 as we would expect from a random normal variable with a mean of 0 and a standard deviation of 10.

Note that for $trend and $random, I used "na.rm=TRUE" to ignore the missing values. This is because the algorithm used to extract the seasonal component of the time series needs about half a cycle (15 observations) on the beginning and end of the sequence to figure out the seasonal variation. For this reason, both the head and tail of $trend and $random contain missing values.

In the final command from the block above, I request a correlation between decOut$random and the original random normal variable that I used to create the time series (being careful to only enter complete observations into the calculation in order to avoid the missing data in decOut$random). The correlation of $r = .83$ is quite large, showing that decompose() was able to extract the irregular component of the time series with a reasonable degree of success, albeit not perfectly. The data in decOut$random would probably be suitable for analysis using the conventional linear techniques that we are used to, correlation, regression, and ANOVA, because the filtered time series very likely now has a property called **stationarity,** that is, it contains no trend component and no cyclical component.

I say that it can "probably" be used and that it is "likely" to be stationary because we do not know for sure how successful the decompose() procedure was in removing the trend and seasonality components from the original time series. To determine this we need some new diagnostics that can help us make better judgments about the properties of the time-series data we are examining. One of the most essential tools in the time-series analyst's toolbox is the *autocorrelation function,* often abbreviated as the ACF. The ACF correlates a variable with itself at a later time period. This concept is easy to understand if you look at Table 11.1.

First, notice that the variable we are going to examine with the ACF is called MyVar. We have seven complete observations of MyVar. Now look one column to the right of MyVar and you will see the first **lag** of MyVar: every data point has been shifted down by one time period. Look all the way to the rightmost column and you will see that every observation of MyVar has been shifted down by two time periods. Now imagine correlating MyVar with each lagged column of MyVar (ignoring missing data). For time-series data that contains a trend or seasonal variation, you will see a pattern of correlations at different lag values. These correlations take on a very distinctive pattern

TABLE 11.1. An Example of Lagged Data to Illustrate Autocorrelation			
Observation #	MyVar	lag(MyVar, k=1)	lag(MyVar, k=2)
1	2		
2	1	2	
3	4	1	2
4	1	4	1
5	5	1	4
6	9	5	1
7	2	9	2

depending on whether there is a trend, seasonality, or both. To make these patterns evident, an ACF plot shows a bar plot of correlation values at different amounts of lag. Let's take a look at a few of these to get a flavor of how to diagnose various kinds of nonstationary data.

First, let's look back at the first time series variable that we created. Because we used set.seed() before, I can now repeat the code right here and get the exact same results to provide a reminder of how we did it:

```
set.seed(1234)
tslen <- 180
ex1 <- rnorm(n=tslen,mean=0,sd=10)    # Make a random variable
acf(ex1)
```

The final command above, acf(ex1), produces an autocorrelation plot that shows a completely stationary process, as shown in Figure 11.6.

The height of each little line in Figure 11.6 shows the sign and magnitude of the correlation of the original variable correlated with itself at different amounts of lag. The first line on the left, which you can see is equal to $r = 1$, shows that the variable is perfectly correlated with itself at zero lags. After that first tall line, all of the other lines are very small in height. Some are negative and some are positive but there is no discernible pattern to this variation. The horizontal dotted lines show the threshold of statistical significance for positive and negative correlations. For a stationary time-series process, all of the lagged correlations (other than zero lag) should be nonsignificant and there should be no pattern to the size of the correlations or to the variations between positive and negative correlations.

Now let's see what happens when we add the trend.

```
tex1 <- ex1 + seq(from=1, to=tslen, by=1)    # Add the fake upward trend
acf(tex1)
```

After adding the positive trend to the random normal variable, as shown in Figure 11.7, we get a very characteristic ACF with strong positive correlations

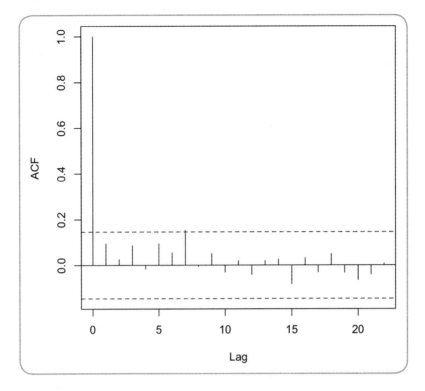

FIGURE 11.6. A stationary variable typically contains no significant lagged correlations.

over many lags. You may want to refresh your memory by looking at the time–series plot of the raw data in Figure 11.2. As always, the correlation of a variable with itself is exactly $r = 1$ (at zero lags). The autocorrelation at one lag is nearly as high, about $r = .98$. This ACF only shows the first 22 lags—an arbitrary choice that can be overridden with the lag.max parameter. The autocorrelation is significant on all 22 of them. Of course, when we created our fake time series variable tex1, we put a superobvious trend in it by using a sequence of integers from 1 to 180: a time series with less obvious trends will have smaller ACFs, particularly at longer lags.

Finally, let's go back to our last fake time series, tex3, which contains both a trend and seasonality. This is how we created that variable:

```
ex3 <- rnorm(n=tslen,mean=0,sd=10)
tex3 <- ex3 + seq(from=1, to=tslen, by=1)      # Add the fake upward trend
tex3 <- tex3 + sin(seq(from=0,to=36,length.out=tslen))*20
acf(tex3)
```

If you examine an ACF for tex3 you will see a very similar ACF to the one in Figure 11.7. Now you may also remember that we have also used the decompose() function to separate tex3 into trend, seasonal, and irregular components

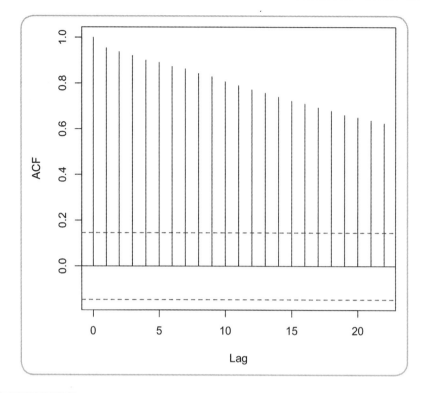

FIGURE 11.7. A time series with a trend shows many positive lagged correlations.

that we can access inside the decOut object. Try running the following command:

```
acf(decOut$trend,na.action=na.pass)
```

Not unexpectedly, the ACF plot of the trend component of decOut will also look very much like Figure 11.7. Note that the na.action=na.pass parameter in the call to acf() gives that function permission to ignore the missing data at the head and tail of decOut$trend. You might notice that the horizontal axis is calibrated very differently from the earlier ACF we examined. Because decOut$trend comes from a decomposition, the horizontal axis is now calibrated according to the frequency = 30 specification we used when we created the time series. Now the rightmost ACF line represents a lag of 22/30 days (about .73 on the horizontal scale of Figure 11.7).

A much more interesting ACF plot comes from the seasonal component of the time-series decomposition. The results of the following line of code appear in Figure 11.8:

```
acf(decOut$seasonal)
```

The seasonal component of a time series, when autocorrelated at different lags, often shows this oscillating pattern of positive and negative correlations. In fact, the strongest negative correlation shown in Figure 11.8 occurs right near 0.5 on the horizontal axis—indicating that half of a cycle occurs in half of a month. This corresponds closely with the roughly 1-month-long oscillation that we put into our fake time series data using a sine wave.

Finally, let's look at the irregular component of the decomposed time series. If the decomposition has been successful, and all of the trend and seasonal components have been removed from decOut$random, we should see an ACF that is very similar to the one in Figure 11.6. Specifically, a stationary time-series process will generally have no statistically significant autocorrelations. Among people who study signal processing, a stationary process is sometimes referred to as "white noise." You can experience an audible form of white noise by tuning a radio to a point on the dial where there is no broadcast: the resulting smooth static is called white noise because it contains a blend of all frequencies, rather than any single dominant frequency. The process of removing trend and cyclical components from a time series is thus also sometimes referred to as *whitening*. The process of whitening a time series is never perfect, so it is possible to occasionally see a small but significant autocorrelation, especially in the earlier lags. What you want to look out for and avoid, however, is a strong pattern of

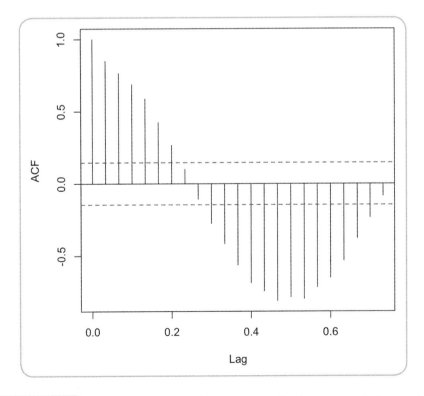

FIGURE 11.8. ACF of the seasonal component of a decomposed time series.

multiple autocorrelations that are significant. Here's how we create an ACF of the random or irregular part of the decomposed time series we created:

```
acf(decOut$random,na.action=na.pass)
```

The resulting ACF in Figure 11.9 shows about nine autocorrelations that are statistically significant (i.e., the height of the bar pokes out above or below the horizontal dotted lines), not counting the ever-present perfect autocorrelation at lag = 0. What is more problematic than these small autocorrelations, however, is the overall pattern. It is evident from the pattern of positive and negative autocorrelations that the sinusoidal pattern we introduced into the fake time series that we created is still present at a low level in these data. Thus, our process of whitening was imperfect. We can also perform an inferential test about whether or not this is a stationary process by using the augmented Dickey–Fuller test, adf.test(), which is in the "tseries" package:

```
install.packages("tseries")
library(tseries)
decComplete <- decOut$random[complete.cases(decOut$random)]
adf.test(decComplete) # Shows significant, so it is stationary
```

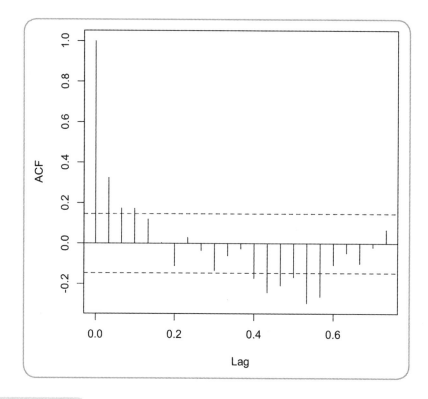

FIGURE 11.9. ACF of the irregular component of a decomposed time series.

The adf.test() procedure in the final line of code above yields the following output:

```
Augmented Dickey-Fuller Test
data: decComplete
Dickey-Fuller = -5.1302, Lag order = 5, p-value = 0.01
alternative hypothesis: stationary
```

The alternative hypothesis for this test is that the process is stationary. Because we rejected the null, we can conclude that this process *is* stationary. As a result, we can feel comfortable proceeding with a substantive analysis that uses the irregular portion from the decomposition even though we had a few autocorrelations that were statistically significant. Although there is still some possibility of finding spurious effects in these data, we have done a reasonably good job of removing trend and seasonal effects.

EXPLORING A TIME SERIES WITH REAL DATA

Now that we are armed with a process for decomposing a time series—which includes removal of the trend and seasonal components—as well as a diagnostic strategy for examining the results, we can take a look at some real data. Let's try examining the built-in data set called EuStockMarkets—recording the daily closing prices of four different European stock markets between 1991 and 1998. In R you can use head() or tail() to review a few observations in this data set. In R-Studio you can also use View() to get a tabular overview of the data. There are four vectors in an R structure known as a multivariate time series (abbreviated "mts"):

```
str(EuStockMarkets)
```

The resulting output appears below:

```
mts [1:1860, 1:4] 1629 1614 1607 1621 1618 . . .
- attr(*, "dimnames")=List of 2
..$ : NULL
..$ : chr [1:4] "DAX" "SMI" "CAC" "FTSE"
- attr(*, "tsp")= num [1:3] 1991 1999 260
- attr(*, "class")= chr [1:3] "mts" "ts" "matrix"
```

You can see that each vector contains 1,860 observations and that the four vectors have the names DAX, SMI, CAC, and FTSE. The "tsp" attribute indicates that the series starts in 1991, ends in 1999, and has 260 observations per unit (stock markets are generally closed on weekends, so the data only account for the days of the year when trading occurs). Let's pose a research question that any investor would enjoy: If I wanted to diversify my holdings in index funds (which are usually created to track a particular stock market), I would want to choose the two funds that are correlated at the lowest level with one another. To explore this question, we can examine which pair of stock markets is least closely correlated with the others, but we will first have to remove the trend and any seasonality that exists. We can begin by looking at all four stock markets together—plot() has an extension that handles multivariate time series elegantly. The following command yields a four-paned plot:

plot(EuStockMarkets)

The upward trend is clear in the plots shown in Figure 11.10. All four of these time series end at a much higher level than they start. That will cause spurious correlations among them. You should run cor(EuStockMarkets) to confirm this. At a minimum, we must remove the trend before proceeding with any substantive analysis. Statisticians have found a simple and effective technique for removing a simple trend from a time-series analysis that does not require a complex decomposition process. The technique is called **differencing** and it goes back to the table of lagged data that I showed earlier in the

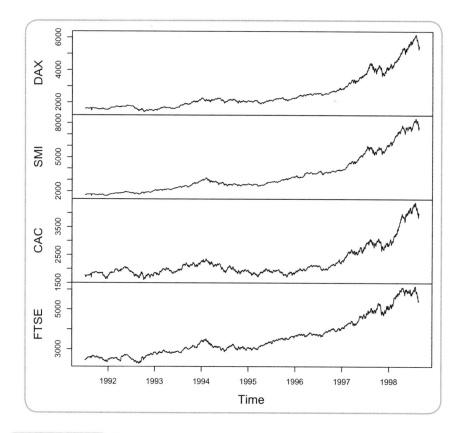

FIGURE 11.10. Time-series plot of four European stock markets.

chapter (Table 11.1). If you subtract the second element in a time series from the first element, you have a difference between the two observations that shows the amount of change between the two observations. This simple method of differencing tends to flatten out any trends that occur over time. The effect is easy to demonstrate using the same strategy that we used to create a trend in our original fake time-series variables:

```
seq_len(10)
```

This code produces the following sequence of ten integers:

```
[1] 1 2 3 4 5 6 7 8 9 10
```

By wrapping that sequence in the diff() command, we can show what happens with the simple one-lag differencing:

```
diff(seq_len(10))
```

The result shows the difference between each neighboring pair of elements:

```
[1] 1 1 1 1 1 1 1 1 1
```

The diff() command created differences at the first lag (hence only nine remaining observations). This simple step of differencing demonstrates how the result is a stationary series (with no variance). Let's apply differencing to the stock market indices and then plot the result:

```
plot(diff(EuStockMarkets))
```

You can see in Figure 11.11 that the trend component has effectively been removed. Note the substantial difference in variability from the early parts of the time series when compared to the later parts. This **heteroscedasticity** may indicate some change in the volatility of the markets and may be of interest, depending upon the research question, but our main goal for now is to see if we effectively removed the trend and created a stationary time series. We can now test for stationarity with the adf.test() procedure. Let's start with DAX:

```
adf.test(diff(EuStockMarkets[,"DAX"]))
```

That adf.test() command produces the following output:

```
    Augmented Dickey-Fuller Test
data: diff(EuStockMarkets[, "DAX"])
Dickey-Fuller = -9.9997, Lag order = 12, p-value = 0.01
alternative hypothesis: stationary
```

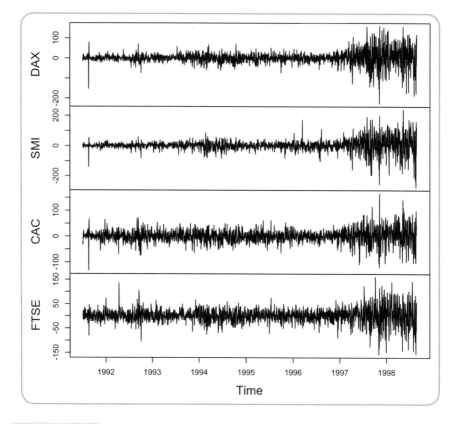

FIGURE 11.11. Time-series plot of four European stock markets after differencing.

Remember that the augmented Dickey–Fuller test examines the null hypothesis of nonstationarity. This test is significant, so we can reject the null and take this test as evidence for stationarity. Using the command acf(diff(EuStockMarkets[,"DAX"])) you can also confirm that the resulting ACF has very few significant autocorrelations.

In conjunction with the significant adf.test() results, we can safely conclude that this time series is stationary. I ran the augmented Dickey–Fuller test on the other three indices and also examined ACF plots for them. In all cases I concluded that differencing had made the time series stationary. Note that differencing generally cannot remove seasonality from a time series. Although stock markets do sometimes have seasonality and other cyclical components in them, if any such component was present in these indices, it was weak enough not to affect our tests of stationarity. At this point we can run a simply bivariate correlation matrix to address our substantive question of which two stock markets are most weakly related to one another. For a basic test we can simply use cor() on the differenced time series to find our results:

```
cor(diff(EuStockMarkets))
```

That command produces a correlation matrix of the first order differences for the four stock markets:

	DAX	SMI	CAC	FTSE
DAX	1.0000000	0.7468422	0.7449335	0.6769647
SMI	0.7468422	1.0000000	0.6414284	0.6169238
CAC	0.7449335	0.6414284	1.0000000	0.6707475
FTSE	0.6769647	0.6169238	0.6707475	1.0000000

This correlation matrix shows that SMI and FTSE are least strongly related, with approximately $r = .62$. That correlation translates into about 38% of variance in common (just square the correlation to find the proportion of shared variance) between the two indices, quite a lot really, but perhaps different enough to provide some diversity when investing in index funds. By the way, if you cross-check this result against the original correlation matrix of "undifferenced" time series, you will find that the conclusion we have drawn here is different from what the original data would have said. This underscores the importance of removing trend and cyclical components from a time series before undertaking a substantive analysis.

FINDING CHANGE POINTS IN TIME SERIES

Once we have mastered the skills of testing whether a trend exists, removing a trend through differencing, and/or removing seasonality and cyclical components, we can begin to ask substantive research questions with time-series data. Although practitioners in business, engineering, and government often model time series in order to forecast future trends, many researchers examine past data to find out whether an intervention or natural event made an impact in the evolution of the time series. A variety of analytical methods exist for exploring this type of question. One of the easiest to understand is called **change-point analysis.** In a change-point analysis, an algorithm searches through the time-series data to detect and document major transitions. Most commonly, such transitions occur in the mean level of the data. For example, let's say that we measure the traffic on a highway over a period of weeks to get a sense of the volume of cars on a particular route. We continue our measurements through a holiday weekend. Taking the earlier data from the "normal" times together with the data from the holiday weekend, we can detect whether the holiday traffic spiked upward beyond what could be expected based on typical variations in traffic volume. Change-point analysis (of means) allows us to document both the point in time when the transition occurred and the change in the mean level of the time series.

We can continue our previous example of European stock markets with the EuStockMarkets data built into R. Hearkening back to Figure 11.10, we know that the four different European stock markets experienced gradual growth during the period from 1991 to 1999, but was there a particular inflection point when the average value of stocks increased substantially? Using the changepoint package, we can explore this question:

```
install.packages("changepoint")
library(changepoint)
DAX <- EuStockMarkets[,"DAX"]
DAXcp <- cpt.mean(DAX)
DAXcp
```

The cpt.mean() function allows us to detect transition points where the mean of a time series changes substantively. Here is the output displayed by the final line of code above:

```
Class 'cpt' : Changepoint Object
    ~~ : S4 class containing 12 slots with names
        date version data.set cpttype method test.stat pen.type pen.value
            minseglen cpts ncpts.max param.est
    ----------
    Created Using changepoint version 2.2.1
    Changepoint type : Change in mean
    Method of analysis : AMOC
    Test Statistic : Normal
    Type of penalty : MBIC with value, 22.585
    Minimum Segment Length : 1
    Maximum no. of cpts : 1
    Changepoint Locations : 1467
```

The first few lines simply document what the output object contains. When we get to "Method of Analysis," we learn that the algorithm has used "AMOC." This stands for "at most one change" and is confirmed a little later in the output with "Maximum no. of cpts : 1." In other words, we have asked the cpt.mean() procedure to look for one and only one shift in the mean of the time series. We can ask the algorithm to look for any number of change points based on our research hypothesis, but in this case I am looking for one big change corresponding to a historical shift in the economies of European countries during the 1990s.

The "Type of Penalty" in the output refers to a mathematical formulation that determines how sensitive the algorithm is to detecting changes. At the most sensitive level, with no penalty for detecting a change, one might have a series like, 100,100,100,100,100,101,101,101,101,101, where the algorithm documents the very small change point at the sixth element. In contrast, with a high penalty for detecting a change, the algorithm will only identify the very largest shifts in the time series. Because we did not specify in the cpt.mean() command that we wanted to control the penalty, a default was used. The default for AMOC is the "Modified Bayesian Information Criterion" abbreviated as

MBIC. In our use of the cpt.mean() function, the choice of MBIC = 22.585 for the penalty is like a statistical line in the sand: Anything that crosses that line is considered a change worth documenting (Zhang & Siegmund, 2007).

Finally, and most important of all, the cpt.mean() detected a change in the mean of the time series at point 1,467 (out of 1,860 points in the time series). This is not a very convenient number to interpret, so let's plot the results with plot(DAXcp,cpt.col="grey",cpt.width=5) to see both the point in time where the change occurred and how big the shift was. The result appears in Figure 11.12.

The change in means is documented by the horizontal grey lines in Figure 11.12. Before the first quarter of 1997, the DAX stock index chugs along at a low level, near 2,000. After the first quarter of 1997, the index suddenly jumps up to a higher level, somewhat over 4,000. Each grey horizontal line represents the mean level of the index across the whole time period covered by the line. Interestingly, the beginning of 1997 corresponded with an enormous boom in European economies generally and in Germany's economy (where the DAX is located) specifically. In this particular time series, it is of course possible to see, just by eye, that there is a substantial rise in the level of the DAX index starting in 1997. In other data sets with more modest increases and decreases, however, the change-point analysis can reveal mean shifts that are not evident

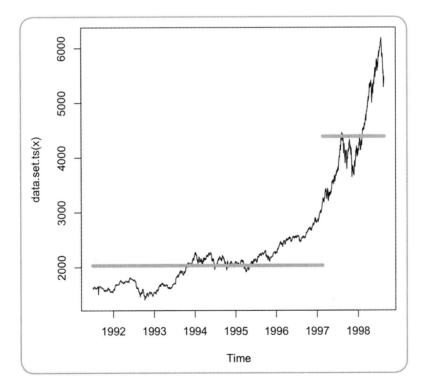

FIGURE 11.12. Change-point analysis of the mean of the DAX stock market series.

to the eye. The cpt.mean() procedure we used can also be harnessed to detect more than one change point in a time series. This enables the possibility of doing field experiments where a phenomenon is measured for a period of time, then an intervention is introduced, and a little later the intervention is taken away. Using change-point analysis, we could detect shifts in the mean of the time series as the intervention is added and removed. By the way, you may have noticed that I did not conduct differencing on the data before conducting the cpt.mean() analysis. Removing these trends from the stock market data would have largely eliminated the growth pattern in the time series, thereby preventing us from finding a shift in the mean. Try it yourself with the command plot(cpt.mean(diff(DAX))).

A parallel analytical procedure, cpt.var() can also be used to detect changes in the variability of a time series over time. You may recall from Figure 11.11 that all four of the stock market indices seemed to hit a period of greater volatility sometime after 1997. By using cpt.var(diff(EuStockMarkets[,"DAX"])), you can produce a change-point analysis of the DAX index. The following excerpt shows the relevant output:

```
Changepoint type : Change in variance
Method of analysis : AMOC
Test Statistic : Normal
Type of penalty : MBIC with value, 22.58338
Minimum Segment Length : 2
Maximum no. of cpts : 1
Changepoint Locations : 1480
```

The major change point occurs at 1480, which we know from Figure 11.12 is part of the way through 1997. This result indicates that just as the market started to increase rapidly in 1997, it also entered a period of more intense volatility, that is, substantially greater variance. You should try the cpt.var() command on the other markets in the EuStockMarkets. You can plot the results to obtain a change-point graph. Note that I analyzed the first order difference scores rather than the raw time-series data to avoid any influence of the trend on the change-point variance analysis.

PROBABILITIES IN CHANGE-POINT ANALYSIS

You might be wondering whether there is a statistical significance test that can be applied to the change-point analysis to understand whether the detected change in mean might have occurred by chance. The cpt.mean() procedure does not conduct a statistical significance test per se, but it does generate a confidence "level" that is expressed as a probability. This confidence level is not the same thing as a confidence interval, but is rather a "strength of belief" about the change in mean (quite Bayesian!). You might interpret it in a similar way as you would an effect size like R-squared in that the value ranges from 0 to 1, and values closer to 1 indicate a stronger effect, and therefore greater surety that the detected change in the mean level of the time series is not due to chance.

There is a little trick to getting cpt.mean() to report the confidence level. Usually, the procedure returns a complex data object that contains lots of fields as well as a copy of the original time-series data. That data object is technically an "S4 class" data object. We can get a much simpler piece of output by setting class = FALSE with the following command:

```
DAXcp <- cpt.mean(DAX,class=FALSE)
DAXcp["conf.value"]
```

Those commands echo the following results to the console:

```
conf.value
    1
```

You can see from this output that the confidence value is 1.0, the strongest possible value. This signifies that our analysis of the DAX data has indeed detected a powerful change in the mean of the time series.

This is another situation where a Bayesian analysis can give us a much clearer view of what is happening in our time-series data than traditional methods. In typical R fashion, there is a package that contains a Bayesian version of change-point analysis. This Bayesian version uses the Markov chain Monte Carlo technique to develop a list of posterior priorities for mean changes *at each point in the time series.* We can examine a plot of these probabilities over time to get a detailed sense of where mean changes are most likely occurring in the time series. The name of this package is bcp, and it works in a similar way to cpt.mean(), but produces a much richer kind of output that we can view in a time plot:

```
install.packages("bcp")
library(bcp)
bcpDAX <- bcp(as.vector(DAX))
plot(bcpDAX)
```

Note in our call to bcp() above that we had to convert the DAX time series to a plain old vector to get bcp() to accept it, using the as.vector() coercion procedure. Plotting the resulting output creates a data display with two corresponding panes, as shown in Figure 11.13. The upper pane shows the original time series and the lower pane shows the probabilities of a mean change at each point in time. You will note that there are isolated spikes that show probabilities near 1 at many points across the timeline. Yet, somewhere near data point 1500 (the beginning of the year 1997 in our original timeline), we see there is substantial density of probability values near 1. We can get a better sense of this by replotting the probabilities with a little programming trick:

```
plot(bcpDAX$posterior.prob >.95)
```

Quick View of ARIMA

One of the most flexible methods of analyzing time series data is a technique known as Auto-Regressive, Integrated, Moving Average, usually abbreviated as ARIMA. Researchers use the ARIMA technique with univariate time series data for two purposes, modeling and forecasting. For modeling, ARIMA can help to show what kind of processes are involved in generating the data. The "AR," or autoregressive component, suggests that past history of a time series can predict its future trends. We saw hints of this when we looked at the auto-correlation function (ACF) and saw a pattern of significant, though diminishing correlations at many lag values. The "I," or integrated component is the opposite of what we accomplished with differencing. In other words, the integrated component of a time series reflects the accumulating influence of growth (or decline) in a series. Finally, the "MA," or moving average component, suggests that prediction errors from recent points can improve prediction of the current point by dampening the influence of random "shocks" on the time series process.

An ARIMA model is often designated as arima(p,d,q), where p is the "order" of the autoregressive component, d is the order of the integrated component, and q is the order of the moving average component. In this case order refers to the number of parameters needed to represent the model, or in the case of d the degree of differencing needed. The process of figuring out the right values for p, d, and q can be quite complex and requires lots of careful work with diagnostic statistical procedures.

Once we have identified the most appropriate values for p, d, and q, we can use ARIMA to construct a prediction model. Such models can perform quite accurately, assuming that the underlying phenomenon that generated the original time series continues in a similar fashion. Developing an appropriate ARIMA model and using it for forecasting also requires several other important steps, such as removing or modeling seasonality and achieving stationarity in the time series. If you are curious, R contains an ARIMA procedure, which you can use to model and predict water levels in the LakeHuron data set:

```
# Run a model with p=1, d=0, and q=1; hold out the last ten values
tsFit <- arima(LakeHuron[1:88], order=c(1,0,1))   # Fit the model
predict(tsFit, n.ahead=10)                         # Predict the next ten values
LakeHuron[89:98]                                   # Compare with the actual values
```

The code above uses the built-in LakeHuron data set to develop an arima(1,0,1) model with the first 88 observations in the time series. I chose p=1, d=0, and q=1 after trying out a number of alternatives and looking at diagnostics. The second line of code predicts the next ten values based on the model parameters that came out of arima(). The predict() function also shows standard errors for each predicted value to indicate the uncertainty around each prediction. Finally, the last line of code

(continued)

shows the final 10 observations of the LakeHuron data set, which we had held back from the analysis process. How good are the predictions?

Strictly speaking, ARIMA is not an inferential technique because there are no hypothesis tests in its essential use. Interestingly, however, statisticians are just starting to apply Bayesian thinking to provide greater flexibility and a richer view of uncertainty around future predictions (for example, take a look at the bsts package). If you want to learn more about ARIMA and the related diagnostic procedures, try searching for "ARIMA tutorial R" to find some good introductory explanations.

This line of code uses a logical test to separate out the low and medium probabilities from the high probabilities. In fact, according to the expression "posterior.prob >.95," every probability value less than or equal to 0.95 gets recoded as FALSE (which in R is equivalent to 0) whereas everything above 0.95 gets coded as TRUE (which in R is equivalent to 1). The display in Figure 11.14 shows that there are two such points just above 500, another two points just above 1000, and then a whole bunch of points just above 1500 with very strong probabilities of being change points—signifying a substantial and

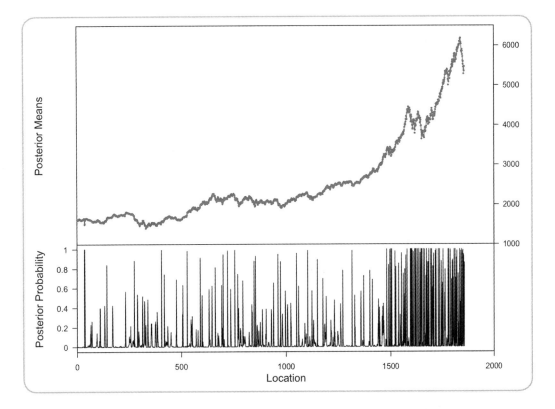

FIGURE 11.13. Bayesian change-point analysis of the DAX stock market.

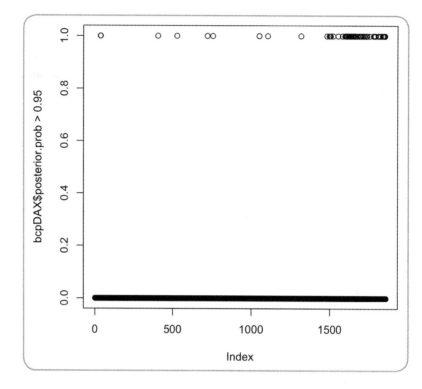

FIGURE 11.14. Plot of the posterior probabilities from Bayesian change-point analysis of DAX, with a split at $p > 0.95$.

sustained change in the mean of the time series. These change points correspond to periods just before 1994 and just after 1996 when there were modest rises in the DAX index. Cross-reference with Figure 11.10 to see what I mean. Most importantly, there is a huge and sustained rise in the mean of the time series early in 1997, just as we detected with cpt.mean(). The density of probability estimates in this time region (i.e., later than 1500 on the X-axis in Figure 11.14) confirm for us that this is a major shift in the mean of the time series.

CONCLUSION

Whereas earlier in the book we analyzed data that did not incorporate considerations of time (so-called cross-sectional data), in this chapter we analyzed data that measured changes over one or more time periods. The big difference from an analytical standpoint is that with cross-sectional data, all observations are independent from each other. In contrast, repeated-measures and time-series data collections provide information on how one or more phenomena change over time, making observations dependent on each other. Repeated-measures research designs tend to measure several different subjects at a few points in time, whereas time-series data tends to focus on one subject over many points in time.

We examined two kinds of repeated-measures analysis: the dependent-measures *t*-test and the repeated-measures ANOVA. The dependent-measures *t*-test has the virtue of simplicity but is limited to examining changes across exactly two points in time. We explored how a paired sample *t*-test is functionally equivalent to analyzing difference scores obtained by subtracting time 1 data from time 2 data. The repeated-measures ANOVA expands our capabilities by providing a method of examining change across two *or more* points in time. For both the dependent-measures *t*-test and the repeated-measures ANOVA we must have multiple cases/subjects where each one is measured at the different time points.

When using repeated-measures ANOVA, we took care to make sure our data were "balanced," meaning that each subject had an observation at each point in time. The traditional calculation techniques used in aov() and other related procedures depend on having balanced data to properly calculate sums of squares and the *F*-test. You can think about the design of a repeated-measures study as a table where each cell is the intersection of a subject and a time period: we must have exactly one observation for each cell or the results of the ANOVA may be confusing or incorrect.

Although there is no strict limit on the number of time periods that can be included in a repeated-measures ANOVA, most research designs only include observations captured at a few time periods. For longer studies where we examine one phenomenon over dozens or hundreds of time periods, a time-series design is more suitable. In a time-series design, each point is an element in a long, ordered sequence and we must therefore be cognizant of trend, seasonality, and cyclicality. We examined procedures to break down a time series into these components when necessary.

"Trend" refers to the systematic growth or decline of a time series. A stock market is an example of a time series that tends to grow over time. We can usually remove a trend, if we need to for analytical purposes, by differencing neighboring points. If today's stock market is at 1,000 and yesterday's was at 999, then the first order difference is 1. "Seasonality" refers to a cycle of change that occurs on a regular basis. For example, people tend to take more vacations in winter and summer and fewer vacations in spring and fall and these are patterns that repeat year after year. Cyclicality is conceptually similar to seasonality, but may not occur on such a regular basis.

When we remove the trend and cyclical components from a time series we are left with data that people call by one of four names: the "whitened" time series (as in "white noise"), the irregular component, the noise component, or the random component. These names all have a misleading connotation, because it is this leftover piece after we take out trend and cyclicality that may be most interesting to us from an analytical standpoint. If we are to conduct any kind of correlational analysis on time-series data (including linear multiple regression), we absolutely must remove the cyclical and trend components first, as these will cause spurious correlations if left in the data during analysis.

Although economists, businesspeople, and engineers sometimes use time-series analysis for forecasting applications, from a research standpoint it is often useful to look back through a historical time series and detect the specific point in time where a change in mean (or variance) occurred. Sometimes these changes can be the result of natural, observed phenomena (such as economic changes in a country) and sometimes they will occur as the result of an inter-vention that we planned (e.g., when we change the content of a web page to see if we get more hits). In either case, change-point analysis gives us the capabil-ity to sort through the natural variations—ups and downs—that tend to occur in a time series, to locate the moments when meaningful change in mean (or variance) occurred. We can then use this information to draw inferences either about the naturally occurring phenomenon or the intervention we planned. We explored two methods of accomplishing change-point analysis, one with a more traditional approach (although not a typical null hypothesis test) and one with a Bayesian approach. As always, both approaches are serviceable and may be complementary, though the Bayesian approach gives us a detailed view of the likelihood of a meaningful change at each point in time.

EXERCISES

1. The following two lines of code produce two samples of $n = 100$ points each:

```
grp1 <-rnorm(100)
grp2 <-grp1+runif(100,max=0.1)
```

These two samples are related, because the calculation of the second one includes the data from the first one. Explain what each line of code accomplishes. Then analyze the two groups using the t.test(). Run the t.test() twice, once with paired=FALSE and once with paired=TRUE. Interpret the output of each t.test() and then explain why the results of these two different tests are so different from each other.

2. Download and library the nlme package and use data ("Blackmore") to activate the Blackmore data set. Inspect the data and create a box plot showing the exercise level at different ages. Run a repeated measures ANOVA to compare exercise levels at ages 8, 10, and 12 using aov(). You can use a command like, myData <-Blackmore[Blackmore$age <=12,], to subset the data. Keeping in mind that the data will need to be bal-anced before you can conduct this analysis, try running a command like this, table(myData$subject,myData$age)), as the starting point for cleaning up the data set.

3. Starting with the EuStockMarket data, use differencing to remove the trend from each of the four time series. Plot the results as shown in this chapter. Then use cpt.var() from the changepoint package to detect the one point in each series where there is a sub-stantial change in the variance of the time series. Contrast these results with the results of cpt.mean() as reported in the chapter. Describe in your own words what happened before and after the change point detected by cpt.var().

4. The AirPassengers built-in data set tracks the number of international air travelers between 1949 and 1960. Use decompose() and plot() to provide a view of the trend, cyclical, and random (irregular) components of this time series. Describe in your own words what each of the four panes contains and explain what is happening with the data in each pane.

5. Given that the AirPassengers data set has a substantial growth trend, use diff() to create a differenced data set. Use plot() to examine and interpret the results of differencing. Use cpt.var() to find the change point in the variability of the differenced time series. Plot the result and describe in your own words what the change point signifies.

6. Use cpt.mean() on the AirPassengers time series. Plot and interpret the results. Compare the change point of the mean that you uncovered in this case to the change point in the variance that you uncovered in Exercise 5. What do these change points suggest about the history of air travel?

7. Find historical information about air travel on the Internet and/or in reference materials that sheds light on the results from Exercises 5 and 6. Write a mini-article (less than 250 words) that interprets your statistical findings from Exercises 5 and 6 in the context of the historical information you found.

8. Use bcp() on the AirPassengers time series. Plot and interpret the results. Make sure to contrast these results with those from Exercise 6.

9. Analyze the sunspot.month built-in data set and create a mini-article (similarly to Exercise 7) with scientific background information, appropriate plots, and statistical results.

## CHAPTER 12

Dealing with Too Many Variables

In all of the material in previous chapters, we have conducted analyses on individual variables. In most cases, each variable was a single vector of measurements, such as the number of kilometers driven, or a calculated measurement, such as GPA (an average across the grade points obtained from a number of individual courses). In real data analysis situations, however, we often have multiple measurements of the same thing. Researchers need to have a meaningful method of combining multiple measurements of the same thing into a single composite measurement.

One of the most common examples of a composite measurement is the multi-item rating scale. These scales often appear on opinion surveys where the researcher wants to obtain a reliable estimate of a respondent's standing on a particular hypothetical construct. For example, if one were interested in measuring how much a person liked cheese, one might ask the person to rate his or her level of agreement with each of the following five statements:

1. "I enjoy eating a grilled cheese sandwich."
2. "I often sprinkle grated cheese on my pasta."
3. "Cheese and crackers makes an excellent snack."
4. "Hard cheeses like cheddar are delicious."
5. "Soft cheeses like brie are delectable."

Once the respondent had completed the survey, you would have five different vectors of ratings—one for each item. If you were developing a predictive model of cheese enjoyment, however, you would not want to have to analyze five different models to predict each of the five different ratings. Instead, you would want to think about a way of combining the five different ratings into a composite or summary measure of cheese enjoyment. This is a "grate" idea (sorry, that was a cheesy pun), but you have to be careful before you combine them to make sure that all five of your ratings are playing on the same team. Whether or not they are playing on the same team, in turn, depends upon the correlations among the ratings. If the individual items are all uncorrelated with each other, and you combine them, you will end up with a noise/nuisance variable that cannot possibly predict (or be predicted by) anything. So the five ratings have to be correlated with one another, but how much? And do they all have to correlate with each other by the same amount? And what if one or two out of the original five are not correlated very much with all the others?

We can examine a correlation matrix to address these questions, but the human brain is not well suited for simultaneously viewing the dozens of correlation coefficients that may be involved or for making good judgments about correlations as a group. **Principal components analysis** is a tool that helps to address this situation by reorganizing the internal correlation structure among a group of measurements and displaying it in a compact way that is easy to interpret. The analysis process to extract principal components in effect creates a set of new variables to represent the input variables. If there are four input variables, there are four principal components to represent them. The neat part is that if there is common variance among the input variables, that common variance gets placed in the first component. Whatever else is left gets put in the second component, and so on, until the last component contains whatever is left over. The benefit of this rearrangement is that the last few components can often be ignored, thus simplifying what used to be several variables into one or two components. Keep in mind that a component is nothing more than a linear combination of the input variables. The principal components analysis procedure figures out a set of **loadings**—similar to regression coefficients— that combine the input variables together so that the first principal component captures most of the common variance among the input variables.

Principal components analysis, and its big brothers, factor analysis and independent components analysis, are together known as exploratory factor analysis techniques. These exploratory factor analysis techniques are critically important in big-data applications for a process known as **dimension reduction.** One of the common difficulties with big data is that there are often more variables than we can practically use in an analysis. Data wranglers need techniques for combining variables into composites—to accomplish that one needs to understand the internal covariance structure within a set of variables. This process is called "dimension reduction" because it takes many dimensions— what we call "variables"—and reduces them to fewer dimensions. Exploratory factor analysis is one of the most basic techniques for dimension reduction. In this chapter we examine principal components analysis as the simplest of those exploratory factor analysis techniques.

Let's begin to explore principal components analysis with a built in data set called "iris." The iris data set contains four different measurements of iris

flowers. A total of 150 different plants were measured. With principal component analysis, we can ask whether these measurements might all have the same
underlying source of variance—for example, the overall size of the iris plant. If
all four measurements are simply minor variations on the overall size of a plant,
we could combine them to get a more reliable index of plant size. If we were
horticulturalists who wanted to find the right combination of soil, water, and
light that led to large plants, we could then use our composite index of flower
size as the dependent variable in an analysis or in model development. Here's an
overview of the structure of the data:

```
str(iris) # What is the data structure for the iris dataset?
```

```
'data.frame': 150 obs. of 5 variables:
$ Sepal.Length: num 5.1 4.9 4.7 4.6 5 5.4 4.6 5 4.4 4.9 ...
$ Sepal.Width : num 3.5 3 3.2 3.1 3.6 3.9 3.4 3.4 2.9 3.1 ...
$ Petal.Length: num 1.4 1.4 1.3 1.5 1.4 1.7 1.4 1.5 1.4 1.5 ...
$ Petal.Width : num 0.2 0.2 0.2 0.2 0.2 0.4 0.3 0.2 0.2 0.1 ...
$ Species : Factor w/ 3 levels "setosa","versicolor",..: 1 1 1 1 1 1 1 1 1 1 ...
```

The iris variables include the length and width of a petal as well as the
length and width of a sepal (a leafy component of a flower that usually lies
underneath the petals). The iris data also contains a factor designator, Species, with three levels. For this principal components analysis we will ignore
this factor. In fact, because the principal() procedure we will be using expects
to receive only numeric variables, let's copy the data frame and remove the
factor:

```
irisN <- subset(iris,select=-Species) # Remove the species designator
str(irisN)
'data.frame': 150 obs. of 4 variables:
$ Sepal.Length: num 5.1 4.9 4.7 4.6 5 5.4 4.6 5 4.4 4.9 ...
$ Sepal.Width : num 3.5 3 3.2 3.1 3.6 3.9 3.4 3.4 2.9 3.1 ...
$ Petal.Length: num 1.4 1.4 1.3 1.5 1.4 1.7 1.4 1.5 1.4 1.5 ...
$ Petal.Width : num 0.2 0.2 0.2 0.2 0.2 0.4 0.3 0.2 0.2 0.1 ...
```

Before getting started with the principal components analysis, let's take a
look at a full correlation matrix for these variables. Note that I have rounded
each correlation to three significant digits to make the matrix more visually
interpretable:

```
round(cor(irisN),digits=3) # Show correlation matrix for the iris data
```

	Sepal.Length	Sepal.Width	Petal.Length	Petal.Width
Sepal.Length	1.000	-0.118	0.872	0.818
Sepal.Width	-0.118	1.000	-0.428	-0.366
Petal.Length	0.872	-0.428	1.000	0.963
Petal.Width	0.818	-0.366	0.963	1.000

Even a small correlation matrix like this can be confusing. Remember that the diagonal is 1 because it represents the correlation between a variable and itself. The triangle below the diagonal is the mirror image of the triangle above the diagonal: most analysts simply ignore the upper triangle. When I look at a correlation with the intention of thinking about principal components analysis, I pay attention to the sign of each correlation and I mentally sort the magnitude of each correlation into negligible ($r < 0.30$), interesting ($0.30 < r < 0.70$), and large ($r > 0.70$). With these rules of thumb in mind, these correlations reveal an interesting structure: Sepal.Length is essentially unrelated to Sepal.Width ($r = -0.118$), but is very strongly and positively related to Petal.Length ($r = 0.872$) and Petal.Width ($r = 0.818$). Sepal.Width is only modestly (and negatively) correlated with Petal.Length ($r = -0.428$) and Petal.Width ($r = -0.366$). Finally, Petal.Length and Petal.Width are very strongly correlated ($r = 0.963$). As we examine the results of the principal components analysis below, keep in mind that Sepal.Width seems to be the "odd man out" in the sense that it is the one measurement that seems somewhat disconnected from whatever biological mechanism may control the other aspects of flower size.

There is a principal components analysis function built into the core of the R installation, but it does not support a key feature of the analysis that I would like to demonstrate, so instead we will use the procedure from an add-in package. The principal components analysis procedure "principal()" is part of the "psych" package, which we will need to load prior to first use. In this code, we place the output of the analysis into a new data object, irisNout:

```
install.packages("psych")
library(psych)
irisNout <- principal(irisN)
irisNout
```

Here's a summary of the results:

```
Principal Components Analysis
Call: principal(r = irisN)
Standardized loadings (pattern matrix) based upon correlation matrix
               PC1      h2      u2      com
Sepal.Length   0.89    0.79    0.208    1
Sepal.Width   -0.46    0.21    0.788    1
Petal.Length   0.99    0.98    0.017    1
Petal.Width    0.96    0.93    0.069    1
               PC1
SS loadings    2.92
Proportion Var 0.73
```

```
Mean item complexity = 1
Test of the hypothesis that 1 component is sufficient.
The root mean square of the residuals (RMSR) is 0.13
with the empirical chi square 28.19 with prob < 7.6e-07
```

Let's take a detailed look at all of these pieces of output. Remember that principal components analysis reorganizes the variance and covariance that it receives from the input variables. By default, the principal() procedure tries to squeeze all of the common variance among the input variables into the first (or first few) principal components. There is a parameter for the principal() procedure called "nfactors=" that controls this, and it defaults to "nfactors=1." In the output above, the PC1 column shows the standardized loading of each input variable onto the first principal component. These loadings are like regression weights or correlation coefficients: How strongly does each input variable connect with this synthesized new variable that we call the first principal component? You should be able to see that Sepal.Length, Petal.Length, and Petal.Width all load very strongly onto the first principal component, but Sepal.Width has a much weaker loading (−0.46) than the other three. The minus sign on the loading is just like a negative regression weight: larger values of Sepal.Width signify smaller values of the principal component and vice versa.

The other columns, h2 (**communality**), u2 (**uniqueness**), and com (an index of complexity; Hofmann, 1978), provide additional information about the contribution of each input variable. *Communality* is the proportion of variance in the input variable explained by the principal component(s). Because we have chosen a one-factor solution here, the communality is simply the square of the loading—more generally, it is the sum of the squared loadings across all of the retained principal components. The *uniqueness* is the "opposite" of the communality, that is, it is the unexplained part of the input variable. So the uniqueness is always 1 minus the communality. Hofmann's complexity index captures how "spread out" a variable's loadings are across the factors: a value closer to 1 means more simplicity, while a value larger than 1 means more complexity. For a one-factor solution, Hofmann's index is always 1, by definition.

Although communality, uniqueness, and complexity/simplicity are useful diagnostics, analysts generally pay the most attention to the loadings. Later in the output, there is a section that shows the sum of squares (SS Loadings) and the proportion of variance accounted for by the principal component (Proportion Var)—the latter quantity is like an *R*-squared value that shows how well the input variables work as a group in "predicting" the principal component. Note for the sake of reference that the row called "SSloadings" is referred to by statisticians as the "eigenvalues." Finally, there is a test of the goodness of fit, or sufficiency of the one principal component at accounting for all of the variance in the four input variables. In this case a significant result means that there is a substantial or meaningful amount of variance that is not accounted for. The 0.73 (Proportion Var) fails to account for enough of the variance in the input variables to be considered a good fit.

Because a one-component solution is not a good fit in this case, we want to continue to explore by trying a two-component solution. The difference between this and the previous call is the use of the "nfactors=2" parameter in the call to principal:

```
irisNout <- principal(irisN, nfactors=2)
irisNout
```

The request for a two-factor solution produces the following output:

Principal Components Analysis
Call: principal(r = irisN, nfactors = 2)
Standardized loadings (pattern matrix) based upon correlation matrix

	PC1	PC2	h2	u2	com
Sepal.Length	0.96	0.05	0.92	0.0774	1.0
Sepal.Width	-0.14	0.98	0.99	0.0091	1.0
Petal.Length	0.94	-0.30	0.98	0.0163	1.2
Petal.Width	0.93	-0.26	0.94	0.0647	1.2

	PC1	PC2
SS loadings	2.70	1.13
Proportion Var	0.68	0.28
Cumulative Var	0.68	0.96
Proportion Explained	0.71	0.29
Cumulative Proportion	0.71	1.00

Mean item complexity = 1.1
Test of the hypothesis that 2 components are sufficient.
The root mean square of the residuals (RMSR) is 0.03
with the empirical chi square 1.72 with prob < NA

These results show that a two-component solution works satisfactorily. In the output just above, the very small "empirical chi-square" (1.72) with "prob < NA" indicates that two principal components together account for virtually all of the variance in the original four variables. Just above, in the row labeled "Proportion Var," we see that component one accounts for 68% of the variance in the input variables, whereas component two accounts for an additional 28%.

At the top of the output, the loadings for "PC1" and "PC2" confirm what we expected before: Sepal.Length (0.96), Petal.Length (0.94), and Petal.Width (0.93) all load mainly on the first component, whereas Sepal.Width has a small negative loading on the first component (−0.14) and a large positive loading on the second component (0.98). Note that Petal.Length and Petal.Width also have small "cross-loadings" on the second component, but you can generally ignore any loading that is lower than (an absolute value of) 0.35.

With respect to our goal of dimension reduction, we can now combine Sepal.Length, Petal.Length, and Petal.Width into a single composite indicator of flower size. Before doing so, it is important to understand the scale on which each item is measured, because you cannot simply add or average things that are on different scales (e.g., height in inches and weight in pounds could not simply be added together or averaged). We need to understand whether the ranges and variances of the three items we want to combine are similar. In the case of a multi-item rating scale on a survey, we generally have each item with the same minimum and maximum. For example, a 5-point rating scale is common, in which case the minimum is usually 1 and the maximum is 5. Similarly, we often find that items rated on a 5-point scale have a standard deviation close to 1, just because of the typical way that people respond to survey items. For any multi-item rating scale that meets these conditions, we can create a composite simply by averaging the items. Another alternative—which is mathematically identical from a covariance perspective—is to sum the items, but this is less common.

In the case of the iris data set, we need to check the ranges of the items before we combine them. We can use the summary() command, like this:

```
summary(irisN)
```

That produces the following output:

Sepal.Length	Sepal.Width	Petal.Length	Petal.Width
Min. :4.300	Min. :2.000	Min. :1.000	Min. :0.100
1st Qu.:5.100	1st Qu.:2.800	1st Qu.:1.600	1st Qu.:0.300
Median :5.800	Median :3.000	Median :4.350	Median :1.300
Mean :5.843	Mean :3.057	Mean :3.758	Mean :1.199
3rd Qu.:6.400	3rd Qu.:3.300	3rd Qu.:5.100	3rd Qu.:1.800
Max. :7.900	Max. :4.400	Max. :6.900	Max. :2.500

The summary() command shows that each item has a substantially different scale range. For example, Sepal.Length goes from 4.3 to 7.9, whereas Petal. Width goes from 0.1 to 2.5. We can overcome this problem by standardizing each variable:

```
irisNS <- scale(irisN)       # standardize each variable
flowerSize <- (irisNS[,1]+ irisNS[,3]+ irisNS[,4])/3    # All except Sepal.Width
length(flowerSize)          # Check the vector
mean(flowerSize)            # Examine the mean
sd(flowerSize)              # And the standard deviation
```

We have constructed a new composite variable, which we call flowerSize, by taking the mean of the three variables mentioned above after standardizing them. The output of the length() command shows that flowerSize has $n = 150$ observations, the same as the original data set. The mean() command shows that the mean of this composite is just fractionally different from 0, which is what you would expect when calculating the mean of several standardized variables. Likewise, the sd() command shows that the composite has a standard deviation very close to 1. As an exercise, you should try correlating flowerSize with the other variables, such as Sepal.Length, to see how strongly the composite connects with them. Before you do that, think about whether you expect a large correlation or a small correlation.

Mean Composites versus Factor Scores

If you dig a little more deeply into the inner workings of principal components analysis, you will learn that the loadings of the items onto the principal components can be used to calculate **component/factor scores**. The terms "component score" and "factor score" tend to be used interchangeably, even though component score applies best to principal components analysis and factor score applies best to other forms of factor analysis. If you remember that the iris data set has $n = 150$ observations in it, we can use the output of the principal components command to obtain two vectors of $n = 150$ scores each, one vector for principal component one and one vector for principal component two:

```
facScore1 <- irisNout$scores[,"RC1"]
facScore2 <- irisNout$scores[,"RC2"]
length(facScore1)
mean(facScore1)
sd(facScore1)
```

That code extracted each of the two components into a new vector, facScore1 and facScore2, and then demonstrated that facScore1 consists of $n = 150$ values with a mean of 0 and a standard deviation of 1. Component and factor scores are always standardized values. Now you might be wondering why we created our own composite, called flowerSize, by averaging the standardized values of Sepal.Length, Petal.Length, and Petal.Width, given that the principal() command had already prepared component/factor scores for us.

There are different views about the use of component/factor scores in the statistical community. DiStefano, Zhu, and Mindrila (2009) provide a very readable overview of the key issues. In a nutshell, the method we used to compute flowerSize—taking the mean of standardized raw scores—works well when the input variables all had similar loadings in the principal components analysis. Don't forget that we standardized the raw variables because they were on different scales. When you have items on the same scale, such as is often the case with survey items, you do not need to standardize before computing the mean of a set of items. The virtue of computing the mean in these cases is that each input variable is given equal weight in the composites. Several researchers have demonstrated that equal weights generalize best to new samples. So if you were creating a composite scale that you planned to use for several survey projects, using equal weights (by computing the mean of items) is likely to provide the best results across all of those studies.

A related virtue is that by computing your composite score based on items you choose (using the principal components results to decide), your resulting composite scores have no influence from the input variables you ignored. In contrast, facScore1, as extracted in the code above, contains an influence from Sepal.Width, which we regarded as disconnected from the other three variables. Now, the loading of Sepal.

(continued)

Width on the first component was very small, just −0.14, so it does not have much influence on the factor score, but in other situations the influence could be large enough to be considered problematic.

Next let's give some thought to situations where the mean composite is a poor choice and the component/factor scores are a better choice. For instance, what if the loadings for Sepal.Length, Petal.Length, and Petal.Width were 0.95, 0.65, and 0.35, respectively? In this case, each of the variables has a quite different contribution to the component/factor score. If we gave them all equal weight, by taking a mean composite, we would be cheating Sepal.Length and giving Petal.Width too much influence. In this case the component/factor score might be a much better choice. Likewise, if we are not concerned with generalizability to new samples, then the component/factor scores are a very reasonable choice.

Let's run one last exercise to put some numbers behind the words:

```
cor(facScore1,flowerSize)
cor(facScore2,flowerSize)
```

The correlation between facScore1, the component scores for the first principal component, and flowerSize, our mean composite, is $r = 0.98$. In other words, the two strategies are producing essentially identical results in this case. The second correlation examines the relationship between the second factor score and our mean composite. We should expect a very low correlation here. If you recall the loadings from the principal components analysis, only Sepal.Width had a meaningful loading on the second factor. So the second factor score consists mainly of Sepal.Width, with a small influence from Petal.Length and Petal.Width. Sepal.Width is the one variable that we left out of our calculation of the mean composite score. This second correlation is $r = -0.18$, quite a small correlation value, attributable to the slight contribution that Petal.Length and Petal.Width made to facScore2. If we were going to use a combination of Sepal.Length, Petal.Length, and Petal.Width in a future study, then the mean composite would serve us well. If we simply wanted a combination of Sepal.Length, Petal.Length, and Petal.Width for one-time use, then facScore1 would do fine.

INTERNAL CONSISTENCY RELIABILITY

Let's examine one more topic that is relevant to composite measures: there is a very commonly used measure of the **internal consistency** of a multi-item scale, known as Cronbach's alpha reliability (Cronbach, 1951). Cronbach's alpha summarizes, on a scale of 0 to 1, the coherency of a set of items with respect to their intercorrelations. As such, it provides an important assessment of whether a composite made of a particular group of items is likely to be useful in later analyses. A group of items with a low Cronbach's alpha mainly consists of noise and will tend not to be useful in subsequent analyses.

Cronbach's alpha reliability is the most commonly used measure of internal consistency across the social sciences. Internal consistency is conceptually similar to what we have been thinking about with principal components analysis in that there is a common, invented dimension to which all of the items in the composite contribute. Alpha consistency is the extent to which the items included in a "test" (which in this context means any composite of two or more individual measurements) correlate with the test as a whole. In the social science literature, an alpha value of 0.70 is considered a bare minimum for any multi-item measure, and 0.80 is greatly preferred as a minimum. These rules of thumb are somewhat flexible, however, based on the length of a scale. In the 1950s, when Cronbach worked on this idea, it was quite common to have a scale with 12, 15, or more than 20 items. The correlations among items in a 15-item scale can be quite modest and you can still get an acceptable level of alpha reliability. In these modern days of brief surveys delivered on mobile devices, it is rare to find a scale consisting of so many items: many contemporary scales only have three to five items. With a brief scale, the correlations among items must be much larger to reach an alpha of 0.80 or more.

Let's put the iris data into an alpha analysis and see what it tells us about our errant Sepal.Width variable. Like the principal() function, the alpha() function is part of the psych package which we have already loaded earlier in the chapter. Note in the analysis below that we are examining the unstandardized version of our iris data set:

```
alpha(irisN,check.keys = TRUE)
```

That call to alpha() produces the following output:

```
Reliability analysis
Call: alpha(x = irisN, check.keys = TRUE)
raw_alpha    std.alpha    G6(smc)    average_r    S/N    ase    mean    sd
   0.81         0.85        0.94        0.59       5.9   0.055    3.9   0.86
lower       alpha       upper    95% confidence boundaries
 0.71        0.81        0.92
```

Reliability if an item is dropped:

	raw_alpha	std.alpha	G6(smc)	average_r	S/N	alpha se
Sepal.Length	0.71	0.81	0.86	0.59	4.2	0.077
Sepal.Width	0.88	0.96	0.96	0.88	22.9	0.064
Petal.Length	0.72	0.70	0.77	0.43	2.3	0.075
Petal.Width	0.68	0.73	0.86	0.47	2.7	0.079

Item statistics

	n	raw.r	std.r	r.cor	r.drop	mean	sd
Sepal.Length	150	0.89	0.84	0.83	0.81	5.8	0.83
Sepal.Width	150	0.46	0.57	0.44	0.35	4.9	0.44
Petal.Length	150	0.99	0.98	1.00	0.98	3.8	1.77
Petal.Width	150	0.96	0.94	0.96	0.94	1.2	0.76

Warning message:
In alpha(irisN, check.keys = TRUE) :
Some items were negatively correlated with total scale and were automatically
     reversed. This is indicated by a negative sign for the variable name.

The warning message at the end of this output is very important. You will see from the output that we included all four of the iris items (despite our knowledge from the principal components analysis that Sepal.Width does not belong). The alpha() procedure is smart enough to know that if a measurement correlates negatively with the overall score, then it needs to be inverted before it can contribute to that overall score. The check.keys=TRUE parameter allowed us to do this automatically. Although the alpha() procedure is not actually creating the composite score for you, it temporarily inverts the sign of the given item, so that the item's contribution to the overall score can be accurately assessed. In this case, sensibly, Sepal.Width was inverted.

The overall alpha reliability, with all four items included (with Sepal.Width inverted) is 0.81. As mentioned earlier, this is an acceptable level of alpha, but could we do better? The answer lies in the middle of the output above under the heading "Reliability if an item is dropped," and in the column called "raw_alpha." For each of our three "good" items, if any one of them was dropped, then alpha reliability suffers a decline. In contrast, if we dropped Sepal.Width, alpha reliability would increase up to 0.88, a very worthwhile gain in reliability. These results back up our decision, based on the original principal components analysis, to leave Sepal.Width out of the formulation of our overall composite score. In reporting this result, we would mention the alpha value of 0.88 in reference to our three-item composite. Because we formulated that composite from standardized scores, it would also be worthwhile to rerun the alpha() analysis on the standardized iris data set.

Neither the principal() procedure nor the alpha() procedure is an inferential statistical method (although principal() did have a goodness-of-fit test). Both of these methods are exploratory and descriptive. As such, we are simply trying to find out how data are structured and to see if there is evidence in favor of forming a composite in order to reduce the number of individual variables in later analysis. With that said, you saw that there is a goodness-of-fit test in the principal() procedure (expressed as a chi-square value and a probability) and there is a confidence interval around our estimate of alpha (look at the line that says 95% confidence boundaries). These tests/outputs can help us make good choices about whether to form a composite and which variables to include in a composite, but keep in mind that there is still a degree of judgment that goes into the process of dimension reduction and scale formation. You will have to try out a number of examples in different data sets to get a solid feel for the process of creating a composite scale. Of course, if you use a scale that was developed and documented by another researcher, you can usually follow the item-combining strategy offered by that researcher. Even in the case of a well-developed, previously published scale, however, it is a good idea to use exploratory factor analysis and alpha reliability to test your assumptions.

ROTATION

When we examined the output of the principal() procedure for the iris data set, the loadings were very easy to interpret. Three of our input variables loaded heavily on the first principal component and one of our variables loaded heavily on our second component. In all cases, the "cross-loadings" were very small—in other words, each variable loaded mostly on one and only one component. We can review these results visually using the plot command:

```
irisNout <- principal(irisN,nfactors=2) # We just ran this earlier in the chapter
plot(irisNout)
```

Figure 12.1 shows the loading plot for the iris data set, using a two-component solution. If you look on the lower right, you will see the loadings for Sepal.Length, Petal.Length, and Petal.Width—each of them near 1 on the X-axis (which is the first principal component) and near 0 on the Y-axis (principal component two). Likewise, the loadings for Sepal.Width are near the upper left of the figure, showing a high loading on component two and a low loading on component one. The plot is basically perfect, from an interpretive standpoint, because it shows a clear picture of the two components and that there are no meaningful cross-loadings. You might ask yourself if we were just lucky, or if there is some analytical magic at work that made the solution turn out so well. In this case it is the latter. We can reveal that bit of magic with the command irisNout$rotation, which reveals an indication of the method of "rotation" that the analysis used. In this case the answer comes back to the console as [1] "varimax."

Rotation is a method of adjusting the loadings to make them as interpretable as possible. The principal() command does a **varimax** rotation by default. Varimax rotates the axes for the loadings to maximize how much variation there is in the item loadings. What does it mean to rotate axes? Hold your right arm out sideways from your body and your left arm straight up in the air. Your right arm is the X-axis and your left arm is the Y-axis. Everything you can see in front of you falls somewhere on the coordinate system defined by those axes. Now keep your arms at right angles to one another, but bend yourself sideways at the waist. Everything you can see in front of you is the same as it was before, but because you have shifted your arm-axes, the coordinates have all changed. Varimax does exactly this kind of rotation, to get the loading coefficients lined up in an ideal way.

It might help to show what the original coordinates looked like. We can do that by requesting rotate="none" when we call the principal procedure, like this:

```
irisNout <- principal(irisN,nfactors=2, rotate="none")
plot(irisNout)
```

The results of the plot() command appear in Figure 12.2. This plot shows the "unrotated" loadings. You will immediately notice that the position of points on these coordinates, particularly item one (Sepal.Length) and item two (Sepal.Width) do not fall near the axes. In fact, item two (Sepal.Width) loads at nearly −0.5 on the first principal component (the X-axis). In the parlance of statisticians, these are not "clean" loadings. These are messy and not nearly as easy to interpret as the loadings shown in Figure 12.1. Now stare at Figure 12.2, stick your arms out and do that thing I described above. If you poke your right arm sideways, stick your left arm up in the air, and bend your body at the waist to the left, you can rotate your axes so that both item one (Sepal.Length) and

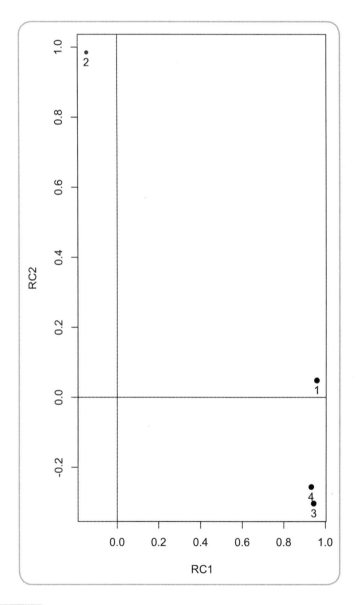

FIGURE 12.1. Loading plot for a two-component solution from the principal() procedure.

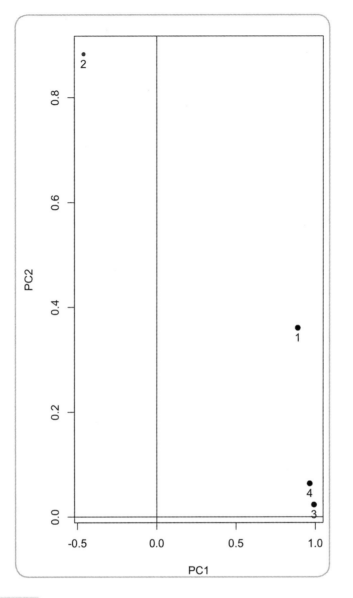

FIGURE 12.2. "Unrotated" loading plot for a two-component solution from the principal() procedure.

item two (Sepal.Width) fall closely on the X- and Y-axes, respectively. When you shift your axes like that, it is varimax rotation.

The name "varimax" refers to the fact that the algorithm tries to maximize the variance of the squared loadings of all of the items. The algorithm keeps twisting the axes until it finds a configuration where the items with the highest loadings fall as close as possible to one axis or the other. When varimax works on more than two principal components, it does the same job in hyperspace. Varimax is an "orthogonal" rotation, which means that the axes are always kept at right angles (90 degrees) to one another. Varimax is the single most popular rotation used in exploratory factor analysis because it is computationally simple to perform, relatively easy to understand, and it generally leads to highly interpretable exploratory factor analysis solutions. This is why the principal() command uses varimax as a default and this is why the first (two-factor) solution that we examined for the iris data set was so "clean" and easy to understand.

Statisticians have developed a variety of other approaches to rotation and you may occasionally find that one of these approaches leads to more interpretable results than varimax. In particular, there are a number of rotation methods, such as "oblimin," that allow the axes to be at an angle other than 90 degrees from each other. This kind of so-called nonorthogonal rotation means that the axes are now "correlated" with each other, adding much complexity to the interpretation. While these can be useful for more complex data sets with larger numbers of items, varimax rotation serves most analysts' needs very well.

CONCLUSION

In this chapter we considered the closely related topics of dimension reduction and composite scale formation. Generally speaking, whenever we have several closely related measurements of the same phenomenon, we want to find a way to combine those measurements into a single composite scale before proceeding with additional analysis. This is particularly true in surveys, where we often have multiple items that tap into the same construct. By using multiple items or measurements to capture a phenomenon, we can actually get a higher degree of accuracy and reproducibility in our measurement processes.

We examined the use of principal components analysis as a strategy for revealing the underlying covariance structure of a set of items. Using the four-item iris data set, we showed how three items were closely related to one another, while a fourth item was the odd man out. This evidence suggested to us that we could combine the three items into a composite scale. While survey items can often be directly combined simply by taking a sum or a mean of the items, in the case of the iris data set the items were on different scales, so we had to standardize them before combining them. We used Cronbach's alpha as a measure of the internal consistency reliability of the resulting scale and found that combining three of the iris items led to a satisfactory level of internal consistency. Similar procedures can be used across a variety of dimension-reduction tasks: use exploratory factor analysis to reveal the structure within a set of items, then choose the items that "go together" and create a composite scale from them. Use Cronbach's alpha to confirm the internal consistency of the set of items in that composite scale.

Finally, we examined the topic of axis rotation. By default, the principal() procedure that we used to conduct principal components analysis subjected the resulting loadings to a varimax rotation in order to make the loadings more interpretable. Varimax maximizes the sum of the squared variances of the loadings of each item on each principal component, and thereby shifts the position of the underlying axes to favor having the strongest loadings fall closely to one axis or another. The resulting mathematical transformations of the loadings tend to make it easier to make sense out of the principal components, and thus show more clearly which items fit together as a group. Varimax is the most popular method of axis rotation and works very well for most data sets.

The tools described in this chapter are, for the most part, not inferential and therefore not used for hypothesis testing. Instead, exploratory factor analysis and Cronbach's alpha can be helpful in creating the variables that will be used later in the analysis process to test hypotheses, using traditional and/or Bayesian methods. The important idea of *parsimony* governs the use of these tools: If we can reasonably reduce the number of variables in an analysis by sensibly combining some groups of variables into composites, then we should.

EXERCISES

1. Use the "?swiss" command to show the help file for the built-in "swiss" data set. Describe in your own words what this data set contains. Speculate, based on the descriptions of the six variables, how a principal components analysis might break down these variables into different groups. Show the output of the str(swiss) command and describe how this confirms the information in the help file.

2. Use the cor() command to examine the correlations among items in the swiss data set. Describe in your own words which pairs of items connect most strongly with one another.

3. Use the principal() command to analyze the swiss data set. Report the loadings of the six items on the first principal component. Which item has the "best" loading and which has the "worst" on the first principal component.

4. Run two-component and three-component analyses on the swiss data set. Which analysis shows a better goodness of fit? Interpret the resulting coefficients and describe one group of three or four items that are likely to fit together well.

5. Conduct an alpha() analysis on the first four items in the swiss data set. Hint: you can use swiss[,1:4] to access the first four columns of the data set. Report the alpha value. Is the alpha value acceptable? Explain why or why not in your own words.

6. Using the results of Exercise 5, report and interpret the confidence interval around the reported alpha value.

7. Run plot() on the output of the two models from Exercise 4. Explain the results in your own words.

8. As you probably remember from earlier in the book, the mtcars built-in data set contains 11 measurements of 32 different cars from the 1970s. Conduct a principal components analysis of mtcars with an eye toward creating one or more composite scales. Report the steps you took and the decisions you made. Report the Cronbach's alpha coefficients for the different combinations of items you considered.

9. Use the USjudgeRatings data set to create a single composite variable representing the favorability ratings of U.S. judges. Report the steps you took and the decisions you made and in particular make sure to justify your choice of items to include and exclude from your composite scale.

All Together Now

Here we are at Chapter 13: you don't believe that is unlucky, do you? By this point I hope that you know that probabilities—and luck—are not tied to any particular numeric value (like 13). Instead, randomness is present in all of the work we do. In particular, the processes of sampling elements from a larger population and of measuring the characteristics of those elements we sample all contain substantial randomness. It is our job as data analysts to model the effects of that randomness. We have to avoid "overinterpreting" results that may look more promising than they really are as a result of random influences. We also have to work hard to find the effects that are present when randomness seems to obscure what we are seeking.

These are the reasons we give so much thought and care to inferential statistical reasoning. We have important decisions to make, we want to make them based on data, but we always want to be alert to the ways in which data can mislead us. The null hypothesis significance test (NHST) is one systematic way to make decisions about data that helps us avoid being misled. In particular, the NHST allows us to choose a level of "Type I error" (false-positive rate) according to convention and tradition and then interpret a variety of observed statistical differences or effects with respect to specific thresholds implied by that choice. This method of statistical reasoning was devised in the 19th and 20th centuries and has been used by researchers for many generations to make decisions based on samples of data.

Lately, however, some fractures have appeared in the foundations of the NHST. Partly, this has come about just from bad habits. When researchers say that this result is marginally significant or that result is highly significant, they are altering the original intent of the method. Although the NHST is not solely to blame, we have seen lots of research over recent years that fails to replicate because the statistical inferences used to document that research were misused.

The other, related problem with the NHST is that it gives us a small and very limited amount of information about the null hypothesis, but only a little information that is directly usable in reasoning about the research hypothesis that we are actually exploring.

For some years now, statisticians and computer scientists have explored a new strategy based on an old idea. An Englishman from the 1700s named Thomas Bayes formulated the idea that we could improve our "posterior" understanding of probabilities by combining our prior understanding with new evidence. This was always a cool idea from the very beginning, but it became more practical as an actual statistical analysis technique as computers and algorithms became more powerful over recent decades. Most of the time, when we are using Bayesian methods, we ask our computers to conduct a kind of "random walk" through a universe of possibilities, looking for those that best fit the available evidence. We saw several examples of such a method known as MCMC—the Markov chain Monte Carlo method of computing a Bayesian solution. All technicalities aside, the great advantage of these techniques is that we can get direct assessments of the probabilities pertaining to our research hypotheses. Now, instead of reasoning indirectly about the null hypothesis, we can go straight for evidence about the "alternative" hypothesis and generate odds ratios and probability distribution information that provides a rich picture of the strength and direction of our results.

We applied both modes of thinking to several of the most basic and important research scenarios: the simple control versus treatment experiment. Our first application was to the t-test, first devised by William Sealy Gosset, AKA "Student," to understand mean differences between two groups. The simplest and most straightforward form of experimentation is to compare the mean of a group of measurements from a control group to the mean of a group of measurements from a treatment group. This situation is where the t-test comes in very handy.

The next step in complexity beyond the t-test was analysis of variance, abbreviated as ANOVA, which gives us the capability to easily compare three or more groups in a single statistical test. Using the simplest form of ANOVA, the "oneway" ANOVA, we can compare a control group to any number of treatment variations. As a group, these variations are referred to as a "factor" where each variant is considered a level of that factor. Later we learned that it is possible to test more than one factor at once. In the classic and most simple of these factorial designs, we might have two factors where each factor has two levels (e.g., male vs. female and right-handed vs. left-handed). More complex designs are possible, of course, but perhaps the most interesting innovation in factorial designs is the ability to test for interactions—those combinations that work together to meaningfully shift the pattern of results.

All of the foregoing tests focused on comparing the means of different groups, but a whole new vista of possibilities opens up when we considered the associations that may exist between variables. When we work with pairs of metric variables, such as height and weight, we can use Pearson's product–moment correlation, one of the essential workhorses of statistics. For categorical variables, we can use a chi-square contingency test, also developed by Karl Pearson. In both cases, we can understand the strength of association between variables and use this to reason about phenomena that may be related to one another.

A step-up in complexity from pairwise associations led us to linear multiple regression, a technique that uses correlations (technically, covariances) as the ingredients in an analysis of how a set of predictors relate to an outcome variable. Although this technique can be used for forecasting, the more common use in research settings is to understand the "strength" of prediction for each independent variable in a set. The magnitude of the regression weights (slopes) in the prediction equation and the inferential tests on those weights can answer complex research questions about a range of interrelated causes and effects. Whenever there are two or more predictors, we learned that it is also possible to test for interaction effects that are analogous to those that can appear in factorial ANOVA.

Of course linear multiple regression works best when the dependent variable (the outcome) is a metric variable such as heart rate, age, or calories. While this covers quite a large number of research situations, in some cases we want to use our predictors to understand a categorical result. In the simplest of these situations, we have a binomial outcome variable (such as correct/incorrect or sick/healthy) that we want to correctly predict using several different independent variables. In this case, we use logistic regression, which models a binary outcome using an inverse logit function (which looks like an S-curve).

We ended our review of inferential techniques by considering data that included considerations of time. These data differ from the independent observations that we examine for the other techniques and as a result must be treated carefully and thoughtfully before analysis. In the end, however, we look into these data for the same kinds of reasoning that we did with non–time-dependent data: changes in means, changes in variability, and associations. We explored two different families of techniques. For a small number of time periods and an arbitrary number of cases or subjects we used the dependent measures *t*-test and the repeated measures ANOVA. When we had measurements from many time intervals, we switched to time-series analysis. With time series, one must explore the data to look for evidence of trends, seasonality, and cyclicality before beginning a substantive consideration of the data. After necessary modifications, we subjected the time-series data to change-point analysis, which allowed us to detect the points in time where substantial changes in means or variances manifested.

Finally, in the most recent chapter we examined what to do in the common instance where we have too many variables. The essential goal here is to understand the internal structure of multiple measurements so that we can perform dimension reduction—that is, choosing and combining variables into a smaller group of composite scales.

THE BIG PICTURE

For each of the statistical techniques we considered there is an ideal context in which that technique can be applied; it is important for the data analyst to know whether the technique is appropriate for the situation. For example, if there are two group means to compare, a *t*-test will do the trick, but if there are three or more, you should probably use ANOVA. Likewise, we generally examined two different strategies for reasoning about the statistical results we observed—the traditional, frequentist method and the Bayesian method. But these decisions do not capture the larger context of what we are trying to accomplish by using inferential techniques.

Data stored in a data set and analyses reported on the R console are useless without a corresponding act of communication. While lots of people (I for one) find it fun to manipulate and "play" with data just for my own amusement, what really matters is what I can say to other people, based on the data I possess and analyze. If I am a researcher, I may communicate my results in an article, whereas if I am working in a business, I may communicate them in a presentation or a report. In each case I am trying to use my data to communicate something about a phenomenon—a problem, a situation, a challenge, a result—to other people with the goal of influencing their thinking or behavior. That is a heavy responsibility! There's a great little book, originally published in 1954 but updated many times, called *How to Lie with Statistics* (Huff, 1993). In this book the author points out the many ways that statistics are misused and misinterpreted—both intentionally and unintentionally influencing people to believe the wrong thing. You do not have to look very far on the Internet or on television to find awful examples of statistics being misused and misinterpreted. You may remember way back at the beginning of the book that I talked about the idea that you cannot prove anything with statistical evidence because of its probabilistic nature. Just do a search on the Internet for "data proves" and see how many writers have violated that rule. It's amazing! And not in a good way!

If you have gotten this far in the book I hope that you are a person of good character and intention and that you do not want to mislead people. Rather, you strive to be seen as credible and persuasive, both authoritative and reliable in your conclusions. To accomplish those noble aims you must take great care with your data and the analytical techniques you use. Taking care with your data and your analytical conclusions is not easy and there is always more to learn. I'm sure you realize that although you spent many weeks slogging through the exercises in this book, you have barely scratched the surface of the vast panoply of analytical techniques that exist out there. But take heart, because you already know enough to be smart and careful.

In particular, you now have many of the ingredients you need to check your own assumptions. Because we used R in this book and because we explored so many different simulations and tests and scenarios and options, you know that you can use the same strategy going forward. Anytime you are using a statistical reasoning technique, whether it is one you learned in this book or one you picked up later, you can, should, and must check your assumptions. Make sure everything makes sense from multiple angles. Use lots of plots and other data displays to make sure that your visualization of the data matches your mental model. Explore your data, deploy your analytical strategy, and then cross-check your results. Ask questions if you don't understand what you are seeing and read more if you don't understand the answers. Only when you are satisfied that you have mastered your data are you ready to communicate your results in order to persuade other people. Remember the awesome responsibility that you have not to lead them astray! You may be interested to know that I myself followed this advice as I wrote this book: there were numerous instances where I could not grasp what the output of an analysis was trying to say. I just kept experimenting and reading and cross-checking until I was satisfied that I was doing it right. And it still may not be perfect! Somewhere in the foregoing chapters there is probably an error—hopefully only a small and meaningless one—and you as the reader of this book must be alert to that possibility. Always use a healthy bit of skepticism combined with careful application of your analytical tools to verify what you think you know.

And one last bit of advice, an echo from days gone by: A long time ago, I had a wonderful and kind mentor, a tall and ancient lady named Patricia Cain Smith. Pat taught me a lot of things about research, but one of the most important things she taught me was how critical it is to start your analysis with clean data. For a lot of the exercises in this book, we worked with the built-in data sets that are packaged into R. This was mainly for your convenience, because I did not want you to get bogged down in the busy work of downloading data, figuring out formats, and fiddling around with data-import options. These built in data sets are neat, but using them largely shielded you from the messy reality that exists in most "wild caught" data sets. Just as I pointed out in the previous paragraph, the importance of checking your assumptions about your analytical results, Pat Smith got me in the habit of checking my basic assumptions about the contents of my data sets. I know it was not her original idea, but she was the first one to convince me that a competent analyst spends 80% of his or her time cleaning the data set before spending the remaining 20% of the time doing the analysis. Sometimes this idea is a bit painful to contemplate because of the large time investment involved, but I have found that they are words to live by.

You will be happy to know that there are no formal exercises for this final chapter, but you do have one more important task ahead of you if you want to put the finishing touches on this phase of your statistical training. Look on the Internet, for example at the *data.gov* website, to find a data set of interest to you. Download and import it into R. Spend some time with the documentation (the metadata), if there is any, or directly poking through the data, if there isn't, and familiarize yourself with what you have before you. Take Pat Smith's advice and check every variable for anomalies. Use lots of histograms, descriptive statistics, scatterplots, and every other tool you can think of to look in the data and make sure that it fits your understanding of what should be there. Fix everything that needs fixing and leave a trail of code and comments that shows what you have accomplished.

Once you are satisfied that the data are as clean as you can possibly make them, practice the analyses you have learned from this book. Almost every data set of any meaningful size contains a grouping variable, some variables that can be used as predictors, and some variables that can be used as outcomes. Challenge yourself to learn everything you can about what these variables have to say. Then write about it, as coherently and convincingly as you can. Last of all, show your work to someone you know, someone who can comment on the persuasiveness and clarity of what you have written.

Most importantly, have fun and keep on learning!

APPENDIX A

Getting Started with R

R is an **open-source** software program for manipulating data, conducting statistical analyses, and creating graphs and other data visualizations. "Open source" means that the program was created by a group of volunteers and made available to the public for free. The first step on the path to using R is to download and install the program onto your computer (unless your school has provided a computer lab where R is already installed: ask your instructor). The volunteers that build and maintain R have created a website for everything related to R (*http://cran.r-project. org*). There is a section on the main page that says something like "Download and Install R" followed by links for each of the three well-known operating systems: Windows, Mac, and Linux. You probably have a Windows or a Mac computer, so click on the appropriate link and follow the instructions written there. For Windows users who are installing R for the first time, there is usually a separate page with additional information and answers to common questions. Most people must download what is called a **binary,** meaning that the software that you download is ready to install and run. You may notice some discussion of **mirrors.** A mirror is simply a copy of a website that is stored in another part of the world. Anytime you have to choose among mirrors, simply choose the one that is geographically closest to your location. Once the binary has been downloaded to your computer, for Windows and Mac users there are just a few clicks needed to install the program and make it ready to run.

Depending upon what kind of computer you have and how experienced you are with installing programs, you may get stuck somewhere in the installation process. Rather than trying to explain every possible solution you may need, let me refer you to some excellent resources that you can consult to get help. First, there are thousands of websites dedicated to helping you install R and solving problems that you may encounter along the way. Try one of the following searches in your favorite search engine:

installing R

how to install R

how to install R on mac

how to install R on windows

troubleshooting R

Alternatively, there are loads of videos on YouTube that walk you through the whole process of installing R on most types of computers. You can use the same searches as shown above—look for the recent videos that also have many views. As you watch some of these videos you will also notice lots of information about installing another program called **R-Studio** as well as information about installing R **packages.** R-Studio is an additional software program that provides an integrated development environment (IDE) for creating R code and working with data. I used it extensively to create the code for this book and preview all of the figures that you see in these pages. Your instructor may want you to work with R-Studio, although it is not required for this book. There is more information about installing R-packages below.

Finally, if you are truly stuck and have run out of new things to try, you may need to communicate with someone about the difficulty you encountered. For this purpose there are several user forums where people ask and answer questions. The questions and answers are archived for future use. Here are five of the most popular user forums:

- *Rbloggers*—As the name implies, *http://www.r-bloggers.com* contains blog posts from people who use R; beyond help with installation, this site also contains many tutorials by statistics and analytics professionals.
- *Stackoverflow*—This is a general-purpose programming support forum that has tagging of questions and answers related to R; go to *http://stackoverflow. com/questions/tagged/r.*
- *Crossvalidated*—This is a companion to the Stackoverflow forum that focuses more directly on statistics and less on programming; go to *http://stats.stack-exchange.com/questions/tagged/r.*
- *R-help Mailing List*—This is a good old-fashioned e-mail list that has been active for many years; search the archive at *http://tolstoy.newcastle.edu.au/R* or sign up as a new user in order to post to the list at *https://stat.ethz.ch/ mailman/listinfo/r-help.*
- *Nabble*—Nabble hosts a user forum that is specifically dedicated to R; search for "nabble R" or go directly to *http://r.789695.n4.nabble.com.*

In all of these cases, make sure to search the archives of previous questions and answers before you post a new question about your specific problem. Active users of these sites prefer not to answer a question that has previously been answered.

RUNNING R AND TYPING COMMANDS

Once you have successfully installed R on your computer, you can run R in the same way you would start any other software application—usually by pointing and clicking on an icon or menu. Assuming everything has been successful in your installation process, the R program will appear as a single window with menus, some icons at the top and a big chunk of text in the middle. Ignore everything except the text for now. When I started the latest version of R on my computer, here's what I got:

```
R version 3.3.0 (2016-05-03)—"Supposedly Educational"
Copyright (C) 2016 The R Foundation for Statistical Computing
Platform: x86_64-apple-darwin13.4.0 (64-bit)

R is free software and comes with ABSOLUTELY NO WARRANTY. You are welcome
    to redistribute it under certain conditions. Type 'license()' or 'licence()' for
    distribution details.

Natural language support but running in an English locale

R is a collaborative project with many contributors. Type 'contributors()' for more
    information and 'citation()' on how to cite R or R packages in publications.

Type 'demo()' for some demos, 'help()' for on-line help, or 'help.start()' for an HTML
    browser interface to help.
Type 'q()' to quit R.
[R.app GUI 1.68 (7202) x86_64-apple-darwin13.4.0]
>
```

That text appears in what is called the **console,** an area where you can type commands and receive simple text-based output. Much of the work we do in this book will involve typing commands into the console and examining the output that R provides. Reading the output shown above from top to bottom, R reports its version number and a funny name that the volunteers came up with for this particular version: "Supposedly Educational." Other recent releases have been called "Smooth Sidewalk" and "Frisbee Sailing." Rumor has it that these names come from various Peanuts cartoons by Charles Schulz. Anyways, it is important to know what version of the program you are running in case you run across a problem that is peculiar to that version. All of the code in this book has been tested on R version 3.3.0 ("Supposedly Educational"), but you will probably be running a newer version of R when you are working with this book, so keep in mind that some of your code may need to be slightly different than how it appears in this book. Likewise your output may not always exactly match what is in the book.

The startup message goes on to tell you that there is no warranty for the software and that there are some commands you can type to get more information. The text also describes the q() command, which is how you close the R software application. Then it displays the version of the graphical user interface (GUI) which will be different for Windows versus Mac. Finally, at the very end you see the "greater than" symbol (**>**) that is the standard **command prompt** for R. The command prompt lets you know that R is ready to receive your next command. You will probably also see a blinking **cursor** right after the command prompt. The cursor is a little vertical line that shows where the next character will be typed. For your first R command, type help("base") and press enter or return.

R will display a separate window that contains the help file for the **base package.** The base package is the most essential ingredient of the R software program. The base package supports many important capabilities such as arithmetic calculations. And speaking of arithmetic, for your next R command, type 2 + 2 at the command prompt. R should respond immediately with this output:

[1] 4

At first glance this looks a little confusing. The digit 1 shown in square brackets ([1]) is simply an item number that R shows in order to help you keep track of more complex output. Obviously when there is just one line of output it is not really necessary, but R reports it anyway. The real answer we are seeking to our 2 + 2 problem is at the end of the line. R correctly displays 4 as the result of that simple calculation. You should try some other arithmetic problems as well, just to get used to using the command line.

In particular, think about this little brainteaser: 2 + 3 * 2. If you just say it in words, it is two plus three times two. Before you hit enter, though, what do you think the answer will be? You might be surprised when you find that R gives the answer as eight. The reason for this is an idea called **operator precedence.** In this case, the multiplication operator (*) has higher precedence than the addition operator (+) and so the multiplication is calculated first, even though the addition comes earlier when reading from left to right. Because of operator precedence, I recommend that you always use parentheses to tell R specifically what you have in mind. For our brainteaser you could either type (2 + 3) * 2 if you wanted the answer to be 10 or 2 + (3 * 2) if you wanted the answer to be 8.

R gives you direct access to data storage where you can place some information that will be used later. At the command line, type x = 99 and then hit enter. R says nothing in response—in this case no news is good news; the lack of response from R means there is no error and that it has stored away the number 99 in a location called x. We can ask R to report what x contains by typing x at the command line and hitting enter. We have used a single equal sign (=) as an **assignment operator,** in other words we have assigned the value 99 to the location x. There is another assignment operator in R that I actually like better, the assignment arrow (<-), because it shows more clearly what is happening. At the command line type x < - 98, hit enter, and then type x at the command line. You will find that storage location x has now taken on the value 98. The assignment arrow shows that the thing on the right (in this case the 98) goes into the thing on the left (in this case x).

But what is x, exactly? Type mode(x) or typeof(x) at the command line and R will report that x is "numeric." No surprise there! Every storage location or data object in R has a mode (computer programmers usually call this a "data type") and numeric is one of the most basic types. Unlike some other languages, R can change the mode of a data object "dynamically," meaning that the object can change modes at any time based on what is stored in it. For example, if you now type x <- "Lucy" and then type mode(x), you will find that R has cheerfully changed the mode of x to "character" without even letting you know. As you work on more complex problems, you should use the mode() or typeof() command freely and frequently, whenever you need to confirm that an R data object is what it is supposed to be. By the way, notice that the quote marks in x <- "Lucy" are the simple up and down quote marks. If you cut and paste text from another source into the R command line, make sure that you use these simple quote marks. The more curly-looking "smart quotes" that some word-processing programs use will confuse R.

Finally, here is one other R feature that new users sometimes find confusing: it is possible to type a command across more than one line. Another way to look at this is that the enter key does not always cause a command to be processed. Try typing the digit 5 and the plus sign at the command line (this, without the quotes: "5 +") and then hit enter. Rather than processing this incomplete command, the R console responds with a plus sign (+), meaning that it is waiting for more input from you. Now type 6 and hit enter. At this point R has a complete command and it dutifully reports the result, which is of course 11. If you ever find yourself in the confusing situation of hitting enter and getting a plus sign (+) when you expected R to process the command, just press the "esc" or Escape key and R will cancel processing of the incomplete command.

INSTALLING PACKAGES

I mentioned above that the "base" package is the most essential R package and it is automatically included as part of the R installation process. There are about 25 other packages that are also included with the installation of R, and most of them contain essential statistical functions that everyone expects and needs to use, for example, calculating the mean and the standard deviation. For this book, however, we will use many packages that do not come with the basic installation of R and that we must download directly from the Internet. As a result, when you are reading the book or working on exercises, it will be really helpful to have an active Internet connection for those situations where we need to download and install a new package.

Most of the time, and on most computers, it is straightforward to download and activate a new package using the install.packages() command and the library() command. The former uses your Internet connection to fetch the additional software needed from one of the mirror servers. The new software is stored at a specific location in the permanent storage of your computer (R knows where it is and normally you do not need to know yourself). The library() command loads the software into working memory from your computer's storage and prepares it for use in R commands. See if you can install a package yourself. Then type the following command at the command line:

install.packages("modeest")

The modeest package provides methods for estimating the statistical mode (the most frequently occurring data value). On my computer, that command produces the following output on the console:

```
trying URL 'http://cran.mtu.edu/bin/macosx/mavericks/contrib/3.3/modeest_2.1.tgz'
Content type 'application/x-gzip' length 154220 bytes (150 KB)
==================================================
downloaded 150 KB
The downloaded binary packages are in
    /var/folders/9y/x80xz84x5_598z2htv1j3bpm0000gn/T//RtmpoWMgug/
    downloaded_packages
```

The output shows that R is attempting to read the package data from a CRAN mirror called *cran.mtu.edu*. The line starting with "Content type" reveals that this is a compressed (zipped) file that is 154,220 bytes long (150 kilobytes; very small by contemporary standards!). A little string of equal signs shows progress as the package is being downloaded from the Internet. Then R confirms that it has downloaded all of the data and it shows the full path name where the data has been stored on my computer. All of this output is "diagnostic," meaning that it is purely informational and can generally be ignored unless there is an error. Occasionally, the output will contain the word "warning" or "error," in which case you will have to pay closer attention to whatever problem may have arisen.

Occasionally, people will have difficulty downloading and installing new packages on their computers. Most of these difficulties relate to either one's Internet connection or the permissions one has to install and/or use new software downloaded from the Internet. If you are using a computer that was provided by school or work, there may be restrictions on what can be downloaded and installed. Fortunately, there is lots of help available to help you solve these kinds of problems. As always, start by looking for relevant information using an Internet search. I found the following searches to be helpful:

R package installation

R package installation troubleshooting

CRAN package installation troubleshooting

By the way, I chose to install the modeest package quite intentionally: we used a mode() command earlier in this appendix that reported the data type of an object rather than the statistical mode. Given that this is a book that discusses statistics, it is a little annoying that the mode() command does not calculate the statistical mode. Fortunately, the mfv() command (mfv stands for most frequent value) in the modeest package provides a method for finding the most frequently occurring value in a data set. Before we can call mfv(), however, we need to load the modeest package into memory by using the library command:

library("modeest")

When I ran this command, I received the following output on the console:

This is package 'modeest' written by P. PONCET.
For a complete list of functions, use 'library(help = "modeest")' or 'help.start().'

This is a friendly message from R indicating the authorship of the modeest package and reminding you how to get help on the package. As practice, try entering the command library(help = "modeest") to see what new commands the modeest package makes available. You might be surprised to learn how many different ways there are to ascertain the statistical mode!

For now, we can try out the simplest method of finding the mode, which is to identify the most frequently occurring value in a data set. As noted above, we can use the mfv() command to accomplish this. Try typing this at the command line:

mfv(c(1,1,1,2,2))

I'm sure you can guess what the result will be before you hit enter. In a list of numbers with three 1's and two 2's, the most frequently occurring value is 1. By the way, the c() command combines or concatenates a group of individual elements into an atomic **vector,** which is one of the essential R data types for containing multiple objects of the same type. You should use mfv() on a bunch of different combinations in addition to c(1,1,1,2,2) to see what you can find out. For example, what if you had the same number of 1's and 2's?

QUITTING, SAVING, AND RESTORING

When you have completed some work in R and you want to close the program, you can either use the menus or the q() command to quit. In either case, R will prompt you whether you want to "Save Workspace Image?" Generally, you should respond to this prompt by clicking "Save." The workspace image contains two important kinds of information that will help you work with R in your next session. First, the workspace contains all of the data objects that you created while working with R. The workspace also saves a history of the most recent commands you have run in R.

Earlier in this appendix we created a data object called x. If you followed along with the examples, x now contains the character string "Lucy" and this would be saved with the workspace. You can find out what data objects are in your workspace by using the ls() command ("ls" is short for "list").

You can step through the most recent commands in your command history by using the up-arrow key to go backward through history and the down-arrow key to go forward through history. In each case, R copies the saved command to the command line at which point you can edit it to make any necessary changes or hit return to rerun the command. On the Mac version of R, in the title bar area of the main R window, there is also an icon that looks like a little striped table. Click on this icon once to display recent command history and double-click on any command to copy it to the command line. The Windows version of R lacks this interface.

After quitting from R, your data workspace and command history are saved to the permanent storage area of your computer, by default to hidden files. The next time you start R, these files are read back in so that you can pick up right where you left off.

CONCLUSION

This appendix provided some guidance on installing and running R, as well as basic command line functions. In addition, I spent some time discussing the installation and use of new packages—additional software that is not included in the basic installation of R. Using the command line to load new packages, get help, and perform basic manipulations of data objects are essential skills for succeeding with this book, so I hope you will take some time to practice them. Please consult Appendix B for additional information on working with data in R.

APPENDIX B

Working with Data Sets in R

If you have not used R before or have not yet installed it on your computer, you should probably consult Appendix A first in order to get up to speed on the essential concepts and skills you need to use the application. In this appendix, I focus on R as a data management tool and show you the most essential aspects of understanding the contents of a data set. Just as with the rest of the book, you will find this material most useful if you do the examples on your computer in R while you are reading.

The most common kind of data set used in statistical analysis takes a "rectangular" shape by virtue of the fact that it contains one or more columns, each of which has the exact same number of rows. The columns may appear in any order. Rows may also appear in any order, unless the rows comprise a time series data set (see Chapter 11 for more about this). By convention, rows are "cases" or "instances" and columns are "attributes" or "variables." For example, age is an attribute of all living things. A data set with an age variable/attribute would have one column dedicated to that variable and could contain cases/instances representing the ages of all of the members of a family. Rectangular data sets follow a set of conventions that you should always try to verify before beginning an analysis:

- Each attribute/variable has a name and no other column in that data set may have the same name. This makes an interesting contrast to spreadsheets, where you could reuse the same column heading as many times as you like.

- Each row refers to one and only one case; good practice dictates that each row should have a cell that contains an identifier, such as a case number, that makes it distinctive from all other rows. Having a case number makes it much easier to trace problems in the data that may appear later on.

- Each cell of the data set (at the intersection of one row and one column) contains just one data element and all of the cells in that column contain data elements of the same type. If your data were originally entered or stored in a spreadsheet, this can become a source of problems: spreadsheets permit users to enter different types of data (such as numbers or character strings) in the same column. In R, the problem will often appear when one finds that a column one expected to be numeric instead turns out to be character data.

- Empty cells contain a special code indicating that they are empty. In R, the marker code for an empty location is known as NA. Other statistical programs use different codes. Some researchers choose a numeric value to represent the absence of valid data. For example, in a column containing age data, a researcher might enter the code 999 to indicate that the age is unknown for that row. Beware of numeric codes for missing data, lest they erroneously become part of the statistical analysis.

DATA FRAMES IN R

A very common and useful data structure in R is known as a **data frame**. The data frame provides a structure for storing rectangular data sets and enforces rules to make sure that they remain rectangular regardless of what manipulations we may apply. In the example below, we will build a data frame from scratch. In practice, it is more common to import a data set into R and store it in a data frame during the process of importing. To start, we will create a new variable that will serve as a case label to uniquely identify each of five cases:

```
caseLabel <- c("A","B","C","D","E")
```

The c() command combines or concatenates a set of values into a list, which in the R world is called a vector. Each entry is placed within double quotes in order to create a character string variable. Commas separating the different values are outside of the double quotes. Once you have typed the line above, remember that you can check the contents by typing the name of the variable on the command line like this:

```
caseLabel
```

Typing the name of a variable on the command line reports the contents of that variable to the console. Next, we create several new variables, each one containing five elements:

```
age <- c(43, 42, 12, 8, 5)
gender <- c("Male","Female","Female","Male","Female")
weight <- c(188,136,83,61,44)
```

Check your work by typing the name of each variable on a blank command line. The age and weight variables are numeric, while the gender variable is specified with character strings. Next, let's create a data frame. In R, a data frame is stored as a list, where each element in the list is a vector. Each vector must be exactly the same length (in this case 5 elements) and each vector also has its own name. The command to make a data frame is as follows:

myFamily <- data.frame(caseLabel, age, gender, weight)

The data.frame() function makes a data frame from the four individual vectors that we previously created. The new data frame object, called myFamily, contains a copy of all of the original data. Type myFamily at the command line to get a report back of what the data frame contains. When you do this, notice that R has put row numbers in front of each row of data. You might wonder why we needed case labels, if R assigns row numbers anyway. The answer is that if we were to sort the data frame into a new order, for example in ascending order of age, the case labels would stay with their respective cases while the row numbers would not. In ascending order of age, case "E" would appear first in the data set and would be designated as row 1.

Next, let's use the str() command to examine the type of "structure" that R used to store the data frame:

str(myFamily)

Here is the resulting output:

'data.frame':   5 obs. of 4 variables:
$ caseLabel: Factor w/ 5 levels "A","B","C","D",..: 1 2 3 4 5
$ age : num 43 42 12 8 5
$ gender : Factor w/ 2 levels "Female","Male": 2 1 1 2 1
$ weight : num 188 136 83 61 44

In the first line of output we have the confirmation that myFamily is a data frame as well as an indication that there are five observations ("obs." which is another word that statisticians use instead of cases or instances) and four variables. After that first line of output, we have four sections that each begin with "$." Each of the four variables has a storage type that is reported by R right after the colon on the line that names the variable.

Note that gender is shown as a **Factor** with two **levels,** meaning that there are two different options for this particular factor: female and male. R assigns a number, starting with 1, to each level of a factor, so every case that is "Female" gets assigned a 1 and every case that is "Male" gets assigned a 2. Because Female comes before Male in the alphabet, Female is the first Factor label, so it gets a 1. In the terminology that R uses, "Factor" refers to a special type of labeling that can be used to identify and organize groups of cases. Importantly, R has automatically converted both caseLabel and gender from their original representation as character strings into factors. R assumes that you are planning to use both caseLabel and gender as categorical variables in your analysis process, as opposed to thinking about them as small chunks of text. You can override this behavior by using the stringsAsFactors=FALSE argument to the data.frame() function. Type "?data. frame" at the command line to get help on the function and an explanation of how stringsAsFactors works.

Next, let's use the summary() function to show some additional descriptive data:

```
summary(myFamily)
```

The output will look something like this:

```
caseLabel  age            gender      weight
A:1        Min. : 5       Female:3    Min. : 44.0
B:1        1st Qu.: 8     Male :2     1st Qu.: 61.0
C:1        Median :12                 Median : 83.0
D:1        Mean :22                   Mean :102.4
E:1        3rd Qu.:42                 3rd Qu.:136.0
           Max. :43                   Max. :188.0
```

The output is organized into columns, one for each of our variables. The output is different depending upon whether we are talking about a factor, such as caseLabel or gender, versus a numeric variable like age or weight. The columns for the Factors list out a few of the Factor names along with the number of occurrences of cases that are coded with that factor. In contrast, for the numeric variables we get five different univariate statistics that summarize the variable. If you have already read Chapter 1, you are probably familiar with all of these statistics already:

- "Min." refers to the **minimum** or lowest value among all the cases. For this data frame, 5 is the lowest age of all of the cases and 44 is the lowest weight.

- "1st Qu." refers to the dividing line that separates the first quartile from the rest of the cases. For example, if we took all the cases and lined them up side by side in order of age we could then divide up the whole into four groups, where each group had the same number of observations in it. Note that if we don't have a number of cases that divides evenly by 4, then the value is an approximation. You might find Table B.1 helpful as a review of quartiles.

- **Median** refers to the value of the case that splits the set of values in half, with half of the cases having higher values and half having lower values. The median is therefore the dividing line that separates the second quartile from the third quartile.

- **Mean**, as described in Chapter 1, is the average of all of the values. For instance, the average age in the family is reported as 22.

TABLE B.1. Overview of Quartiles			
1st Quartile	**2nd Quartile**	**3rd Quartile**	**4th Quartile**
25% of cases here: These cases have the smallest values.	Next largest 25% of cases here.	Next largest 25% of cases here.	25% of cases here: These cases have the largest values.
First and second quartile together account for every case with a value below the median.		Third and fourth quartile together account for every case with a value above the median.	

- "3rd Qu." is the third quartile. This is the third and final dividing line that splits up all of the cases into four equal-sized parts.

- "Max" is the **maximum** value. For example, in this data frame the highest age is 43 and the highest weight is 188.

Finally, let's look at how to access the stored variables in our new data frame. R stores the data frame as a list of vectors and we can use the name of the data frame together with the name of a vector to refer to each one using the "**$**" to connect the two labels like this:

myFamily$age

Why did we go to the trouble of typing out that long variable name with the $ in the middle, when we could have just referred to the "age" variable as we did earlier when we were setting up the data? When we created the myFamily data frame, we *copied* all of the information from the individual vectors that we had before into a new storage space. So now that we have created the myFamily data frame, myFamily$age refers to a completely separate vector of values than age. Prove this to yourself by appending a new value to the original vector. Then type "age" at the command line to see how it looks:

age <- c(age, 11)
age

[1] 43 42 12 8 5 11

Now verify that this **is not** what is stored in myFamily$age by typing that at the command line. You should be able to confirm that myFamily$age does not contain the 11 at the end. This shows that the data frame and its component columns/vectors is now a separate piece of data. We must be very careful, if we establish a data frame to use for subsequent analysis, that we don't make a mistake and keep using some of the original data from which we assembled the data frame.

Here's a question to strengthen your learning. What if we tried to add on a new piece of data on the end of one of the variables in the data frame? In other words, what if we tried something like this command:

myFamily$age <-c(myFamily$age, 11)

Try it out and see what happens. The resulting error message helps to illuminate how R approaches situations like this. Before we close out this section, let's review the R commands we used in this section:

- c() is the combine/concatenate command that allows you to make a vector from a set of values.
- data.frame() creates a new data frame from a supplied list of individual vectors; the result is a rectangular data structure.
- str() reveals the structure of a data object including the data type of each vector within a data frame.

- summary() summarizes the contents of a data object; in the case of a data frame, it lists each column and provides summary statistics.

- $ is a subsetting symbol that chooses one column from a data frame; for example, myFamily$age chooses the age column from the myFamily data frame.

READING INTO DATA FRAMES FROM EXTERNAL FILES

R offers a variety of methods of connecting with external data sources. The easiest strategy for getting data into R is to use the data import dialog in R-Studio. If you have not used R-Studio before, download and install it on your computer before proceeding. In the upper right-hand pane of R-Studio, the "Workspace" shows currently available data objects, but also has a set of buttons at the top for managing the workspace. One of the choices there is the "Import Dataset" button. This button enables a set of dialogs and options for choosing a delimited text file on your computer to import into R. Two of the most common types of text file are comma-delimited and tab-delimited. Each type of file includes one line of text for each row of data, with quotes around string variables, and a comma or a tab to separate each value.

Files containing comma- or tab-delimited data are ubiquitous across the Internet, but sometimes we also need to access binary files in other formats. The term binary is used here to indicate that the data are stored in a form that is not human readable, but that is highly compact, while at the same time being proprietary to the program that stored it. Commercial statistical programs such as SPSS and SAS have binary file formats that are intended for the use of that program. If you have access to the software that created a binary file, you might find it easiest simply to export the data from that program into a comma-delimited file. For example, if you have data stored in Excel, then you could use the File menu in Excel to "Save As" and export the data to a "CSV"-format file (comma-separated variable). If you do not have access to the necessary software, there are a variety of R packages that one might use to access binary data directly from within R. A comprehensive list appears here:

http://cran.r-project.org/doc/manuals/R-data.html

This page shows a range of methods for obtaining data from a wide variety of programs and formats. Note that for the purposes of this book, we will largely use data sets that are built into the basic installation of R or that are included in add-on packages that we install. If you are using this book as part of a class, additional data sets, in comma delimited format, can be found on the companion website or may be provided by your instructor.

APPENDIX C

Using dplyr with Data Frames

Data frames provide an efficient way to store and access rectangular data sets, but the syntax that we must use to control the variables in each data set sometimes seems clunky and verbose. Fortunately, one of the better-known contributors to R, Hadley Wickham, created a package called dplyr that simplifies the manipulation of data frames. The dplyr package uses a consistent grammar to manipulate the rows and columns of a data frame. To illustrate its capabilities, let's use a data set that comes preloaded with R:

```
install.packages("dplyr")
library(dplyr)
data()
```

That final command, data(), will provide a list of all of the data sets that R has ready for immediate use. For this exercise, let's use the EuStockMarkets data set. Whenever you work with a data frame, it is always important to leave the original data pristine, so that you can go back to the original source if you ever need to do so. So let's make a copy of this data set:

```
euStocks <- tbl_df(data.frame(EuStockMarkets))
euStocks
```

The original EuStocks is stored as a matrix, so we begin by **coercing** it to a data frame, and then we coerce to a tbl_df, which is the native data format for the dplyr package. "Coercion" is a piece of vocabulary from computer science that simply means that we change one data type into another data type. One nice aspect of tbl_df objects is that you can automatically report them to the console in a summarized form. The second command above demonstrates this capability. The output from typing the name euStocks at the command line shows four variables,

DAX, SMI, CAC, and FTSE, each of which is a different European stock index. Let's begin by sorting the data set based on DAX data:

```
arrange(euStocks, DAX)
euStocks
```

If you observe the output carefully, you will see that the output of the arrange() function and the subsequent report of the euStocks object do not match. As is usually the case with R, we must assign the output of the arrange() function to a variable in order for the sorting that it performs to be preserved. To save ourselves a proliferation of different data sets, we will assign the output of arrange() right back to our euStocks data object:

```
euStocks <- arrange(euStocks, DAX)
euStocks
```

The output of the second command shows that we saved the sorted result back to the euStocks object, thus preserving the sort of the data that we requested. We can also reverse the sort using the desc() modifier (meaning "descending") and we can include other variables in the sort:

```
euStocks <- arrange(euStocks, desc(DAX), desc(SMI))
euStocks
```

With the first command above, we sort in descending order, with the primary sorting key as DAX and the secondary key as SMI. These extra elements in the command line are often referred to as "parameters." Parameters allow us to control and modify the basic functioning of a command such as arrange().

The second command above produces a summary of the data frame that you will see reflects the new order of sorting. As we have done before, we assigned the output of the arrange() command back into the euStocks variable in order to avoid creating lots of new copies of our data set.

In addition to the arrange() function, dplyr offers a select() function that allows you to choose a subset of the variables in the data frame. Working with only the variables you need can speed up and simplify the manipulation of a data frame:

```
euStocks <- select(euStocks, DAX, FTSE)
euStocks
```

Now our euStocks data frame consists of just two variables, the DAX and FTSE indices. We can also add calculated variables to the data frame using the mutate command. Here, we create a new column that contains an average of the DAX and FTSE indices:

```
euStocks <- mutate(euStocks, avindex = (DAX + FTSE)/2)
euStocks
```

Note that the data frame now contains three columns, where the third column is an average, calculated from the DAX and FTSE columns. One last function to consider is filter(), which gives the capability of choosing a subset of the rows.

For example, if we wanted only to see those rows where the value of avindex was lower than its mean, we could use the following code:

```
filter(euStocks, avindex < mean(avindex))
```

The dplyr package offers a variety of other functions, such as group_by() and summarize(), which can be very useful for producing tabular reports on the data in a data frame. The package also allows for combinations of commands that can perform powerful manipulations of rectangular data sets with a simple series of commands. Almost everything that dplyr does can be accomplished by other R commands, but in many cases those commands are much more complex and considerably more difficult to understand. By the way, if you have ever studied the topic of relational databases you may realize that the dplyr package uses similar terminology to structured query language (e.g., "select").

As we close out this section, let's again review the new R commands we have mastered. Definitely copy these into your notes as well:

- arrange() is a command that is available when the dplyr package is in use; it sorts a rectangular data set according to the contents of one or more columns; the subcommand desc() can be used to request a descending sort order.

- data() shows the built-in data sets that are available for use; when an argument is supplied, it can also be used to load a data set.

- desc() is a subcommand of arrange() that can be used to request a descending sort order.

- filter() gives the capability of choosing a subset of the cases (rows) based on some criteria that you supply such as a minimum or maximum value.

- install.packages() brings in some new R capabilities from the code repository on the web.

- library() makes the code that was imported with install.packages() readily available for use in the workspace of R.

- mean() is the statistical command for calculating the arithmetic mean.

- mutate() makes it possible to add new variables to a data frame that are calculated from the existing variables.

- select() allows choosing a subset of columns from the complete data frame; compare with filter() that allows choosing a subset of rows.

- tbl_df() converts a regular data frame into a data frame that can be manipulated by the commands in the dplyr package.

References

Aguinis, H. (1995). Statistical power with moderated multiple regression in management research. *Journal of Management, 21,* 1141–1158.

Aguinis, H., Werner, S., Abbott, J. L., Angert, C., Park, J. H., & Kohlhausen, D. (2010). Customer-centric science: Reporting significant research results with rigor, relevance, and practical impact in mind. *Organizational Research Methods, 13,* 515–539.

Akaike, H. (1974). A new look at the statistical model identification. *IEEE Transactions on Automatic Control, 19*(6), 716–723.

Anderson, E. (1935). The irises of the Gaspe Peninsula. *Bulletin of the American Iris Society, 59,* 2–5.

Armstrong, S. A., & Henson, R. K. (2004). Statistical and practical significance in the IJPT: A research review from 1993–2003. *International Journal of Play Therapy, 13*(2), 9–30.

Bellhouse, D. R. (2004). The Reverend Thomas Bayes, FRS: A biography to celebrate the tercentenary of his birth. *Statistical Science, 19*(1), 3–43.

Bohannon, J. (2015). Many psychology papers fail replication test. *Science, 349*(6251), 910–911.

Cadwell, J. H. (1954). The statistical treatment of mean deviation. *Biometrika, 41,* 12–18.

Cohen, J. (1988). *Statistical power analysis for the behavioral sciences* (2nd ed.). Hillsdale, NJ: Erlbaum.

Cohen, J. (1992). A power primer. *Psychological Bulletin, 112*(1), 155–159.

Cronbach, L. J. (1951). Coefficient alpha and the internal structure of tests. *Psychometrika, 16*(3), 297–334.

Daniel, L. G. (1998). Statistical significance testing: A historical overview of misuse and misinterpretation with implications for the editorial policies of educational journals. *Research in the Schools, 5*(2), 23–32.

DiStefano, C., Zhu, M., & Mindrila, D. (2009). Understanding and using factor scores: Considerations for the applied researcher. *Practical Assessment, Research and Evaluation, 14*(20), 1–11.

Efron, B. (2013). Bayes' theorem in the 21st century. *Science, 340*(6137), 1177–1178.

Fienberg, S. E. (1992). A brief history of statistics in three and one-half chapters: A review essay. *Statistical Science, 7,* 208–225.

Fischer, H. (2010). *A history of the central limit theorem: From classical to modern probability theory.* New York: Springer.

Fisher, R. A. (1925). *Statistical methods for research workers.* Edinburgh: Oliver & Boyd.

Gigerenzer, G. (2002). *Calculated risks: How to know when numbers deceive you.* New York: Simon & Schuster.

Gorroochurn, P. (2012). Some laws and problems of classical probability and how Cardano anticipated them. *Chance, 25*(4), 13–20.

Gosset, W. G. ("Student"). (1908). The probable error of a mean. *Biometrika, 6*(1), 1–25.

Greenhouse, S. W., & Geisser, S. (1959). On methods in the analysis of profile data. *Psychometrika, 24,* 95–112.

Haller, H., & Krauss, S. (2002). Misinterpretations of significance: A problem students share with their teachers. *Methods of Psychological Research, 7*(1), 1–20.

Hertz, S. (2001). Ladislaus von Bortkiewicz. In C. C. Heyde & E. Seneta (Eds.), *Statisticians of the centuries* (pp. 273–277). New York: Springer-Verlag.

Hofmann, R. J. (1978). Complexity and simplicity as objective indices descriptive of factor solutions. *Multivariate Behavioral Research, 13,* 247–250.

Hornik, K. (2012). The comprehensive R archive network. *Wiley Interdisciplinary Reviews: Computational Statistics, 4*(4), 394–398.

Huff, D. (1993). *How to lie with statistics.* New York: Norton.

Huynh, H., & Feldt, L. S. (1976). Estimation of the Box correction for degrees of freedom from sample data in randomised block and split-plot designs. *Journal of Educational Statistics, 1,* 69–82.

Jamil, T., Ly, A., Morey, R. D., Love, J., Marsman, M., & Wagenmakers, E. J. (2016). Default "Gunel and Dickey" Bayes factors for contingency tables. *Behavior Research Methods, 48,* 1–16.

Jeffreys, H. (1998). *The theory of probability.* Oxford, UK: Oxford University Press.

Kass, R. E., & Raftery, A. E. (1995). Bayes factors. *Journal of the American Statistical Association, 90*(430), 773–795.

Kruschke, J. K. (2013). Bayesian estimation supersedes the *t* test. *Journal of Experimental Psychology: General, 142*(2), 573–603.

Lawrence, M. A. (2013). *ez: Easy analysis and visualization of factorial experiments.* R package version 3.2. Downloaded from *https://cran.r-project.org.*

Lehmann, E. L. (2012). "Student" and small-sample theory. In *Selected works of E. L. Lehmann* (pp. 1001–1008). New York: Springer.

Magnello, M. E. (2009). Karl Pearson and the establishment of mathematical statistics. *International Statistical Review, 77*(1), 3–29.

Martin, A. D., Quinn, K. M., & Park, J. H. (2011). MCMCpack: Markov chain Monte Carlo in R. *Journal of Statistical Software, 42*(9), 1–21.

Meehl, P. E. (1978). Theoretical risks and tabular asterisks: Sir Karl, Sir Ronald, and the slow progress of soft psychology. *Journal of Consulting and Clinical Psychology, 46*(4), 806–834.

Meredith, M., & Kruschke, J. K. (2015). Package "BEST" (Version 0.4.0). Available at *https://cran.r-project.org/web/packages/BEST/index.html.*

Micceri, T. (1989). The unicorn, the normal curve, and other improbable creatures. *Psychological Bulletin, 105*(1), 156–166.

Morey, R. D., Rouder, J. N., & Jamil, T. (2013). BayesFactor: Computation of

Bayes factors for common designs [Computer software manual]. Available at *http://cran.r-project.org/package=BayesFactor*.

Nguyen, D. T., Kim, E. S., Rodriguez de Gil, P., Kellermann, A., Chen, Y. H., Kromrey, J. D., et al. (2016). Parametric tests for two population means under normal and non-normal distributions. *Journal of Modern Applied Statistical Methods, 16*(1), 141–159.

Nuzzo, R. (2014). Statistical errors. *Nature, 506*(7487), 150–152.

Plackett, R. L. (1983). Karl Pearson and the chi-squared test. *International Statistical Review, 51,* 59–72.

Plackett, R. L., & Barnard, G. A. (1990). *Student: A statistical biography of William Sealy Gosset: Based on writings by Egon Sharpe Pearson.* Oxford, UK: Clarendon Press.

R Core Team. (2016). R: A language and environment for statistical computing. [Computer software]. Available at *www.R-project.org*.

Rouder, J. N., & Morey, R. D. (2012). Default Bayes factors for model selection in regression. *Multivariate Behavioral Research, 47*(6), 877–903.

Rouder, J. N., Morey, R. D., Speckman, P. L., & Province, J. M. (2012). Default Bayes factors for ANOVA designs. *Journal of Mathematical Psychology, 56*(5), 356–374.

Rouder, J. N., Speckman, P. L., Sun, D., Morey, R. D., & Iverson, G. (2009). Bayesian t tests for accepting and rejecting the null hypothesis. *Psychonomic Bulletin and Review, 16,* 225–237.

Sawilowsky, S. S., & Blair, R. C. (1992). A more realistic look at the robustness and Type II error properties of the t test to departures from population normality. *Psychological Bulletin, 111*(2), 352–360.

Smith, T. J., & McKenna, C. M. (2012). An examination of ordinal regression goodness-of-fit indices under varied sample conditions and link functions. *General Linear Model Journal, 38*(1), 1–7.

Snow, J. (1855). *On the mode of communication of cholera.* London: Churchill.

Starkweather, J., & Moske, A. K. (2011). *Multinomial logistic regression.* Retrieved July 10, 2016, from *www.unt.edu/rss/class/Jon/Benchmarks/MLR_JDS_Aug2011.pdf*.

Student. (1908). The probable error of a mean. *Biometrika, 6*(1), 1–25.

Wasserstein, R. L., & Lazar, N. A. (2016). The ASA's statement on *p*-values: Context, process, and purpose. *American Statistician, 70*(2), 129–133.

Welch, B. L. (1947). The generalization of "student's" problem when several different population variances are involved. *Biometrika, 34,* 28–35.

Zellner, A., & Siow, A. (1980). Posterior odds for selected regression hypotheses. In J. M. Bernardo, M. H. DeGroot, D. V. Lindley, & A. F. M. Smith (Eds.), *Bayesian statistics: Proceedings of the First International Meeting* (pp. 585–647). Valencia, Spain: University of Valencia Press.

Zhang, N. R., & Siegmund, D. O. (2007). A modified Bayes information criterion with applications to the analysis of comparative genomic hybridization data. *Biometrics, 63*(1), 22–32.

Index

About the Author

Jeffrey M. Stanton, PhD, is Associate Provost for Academic Affairs and Professor in the School of Information Studies at Syracuse University. Dr. Stanton's interests center on research methods, psychometrics, and statistics, with a particular focus on self-report techniques, such as surveys. He has conducted research on a variety of substantive topics in organizational psychology, including the interactions of people and technology in institutional contexts. He is the author of numerous scholarly articles and several books, including *Information Nation: Education and Careers in the Emerging Information Professions* and *The Visible Employee: Using Workplace Monitoring and Surveillance to Protect Information Assets—Without Compromising Employee Privacy or Trust*. Dr. Stanton's background also includes more than a decade of experience in business, both in established firms and startup companies.